PENGUIN BOOKS

THE SECRET LIFE OF THE GROWN-UP BRAIN

Barbara Strauch is health and medical science editor and a deputy science editor at *The New York Times*. She previously covered science and medical issues in Boston and Houston and directed Pulitzer Prize-winning journalism at Newsday. *The Secret Life of the Grown-Up Brain* is a *New York Times* bestseller.

The Secret Life of the
Grown-up Brain

The Secret Life of the Grown-up Brain

Discover the Surprising Talents
of the Middle-Aged Mind

Barbara Strauch

PENGUIN BOOKS

PENGUIN BOOKS

Published by the Penguin Group
Penguin Books Ltd, 80 Strand, London WC2R 0RL, England
Penguin Group (USA) Inc., 375 Hudson Street, New York, New York 10014, USA
Penguin Group (Canada), 90 Eglinton Avenue East, Suite 700, Toronto, Ontario,
Canada M4P 2Y3 (a division of Pearson Penguin Canada Inc.)
Penguin Ireland, 25 St Stephen's Green, Dublin 2, Ireland
(a division of Penguin Books Ltd)
Penguin Group (Australia), 250 Camberwell Road,
Camberwell, Victoria 3124, Australia (a division of Pearson Australia Group Pty Ltd)
Penguin Books India Pvt Ltd, 11 Community Centre,
Panchsheel Park, New Delhi – 110 017, India
Penguin Group (NZ), 67 Apollo Drive, Rosedale, Auckland 0632, New Zealand
(a division of Pearson New Zealand Ltd)
Penguin Books (South Africa) (Pty) Ltd, 24 Sturdee Avenue,
Rosebank, Johannesburg 2196, South Africa

Penguin Books Ltd, Registered Offices: 80 Strand, London WC2R 0RL, England

www.penguin.com

First published in the United States of America by Viking Penguin,
a member of Penguin Group (USA) Inc. 2010
First published in Great Britain by Penguin Books 2011
2

Printed in England by Clays Ltd, St Ives plc

ISBN: 978-0-241-95307-5

www.greenpenguin.co.uk

Penguin Books is committed to a sustainable
future for our business, our readers and our
planet. This book is made from paper certified
by the Forest Stewardship Council.

To my family

Contents

Part Three: Healthier Brains

Introduction

The Changing Landscape of Middle Age

For most of human history, middle age has been largely ignored. Birth, youth, old age, death have all been given their due. But middle age has not only been neglected, it's not even been considered a distinct entity.

For most of human history, of course, such neglect made perfect sense. Lives were brutal and brief; there wasn't time for a middle. By the time of the Greeks, there was a reverence for maturity; Greek citizens could not become jury members until age fifty, for instance. But a Greek middle age was not even close to our current version. Not that many Greeks made it that far, for one thing—the average life expectancy in ancient Greece was thirty years old. For those lucky souls who lived longer, it was more like reaching a high peak, taking a sniff of the bracing mountain air, and then quickly descending into the valley of old age.

Now, of course, all that has changed. With human life spans stretching out—the average life span in the developed world just a century ago was about forty-seven years and is now about seventy-eight—we have a long expanse of time in the middle when we're no longer chasing toddlers and not yet rolling down corridors in wheelchairs. With that shift, middle age has come into its own. Books have been written, movies made, studies launched.

But even with this newfound attention, one aspect of middle age has remained neglected—our brains. Even as science began to pay attention to what was happening to our bodies and our lives in the middle years, it did not think about what was taking place inside our heads. The prevailing view was that a brain during midlife was, if anything, simply a young brain slowly closing down.

Now that's changed, too. With new tools such as brain scanners, genetic analysis, and more sophisticated long-term studies, the middle-aged brain is finally getting its due. Much of the new attention, to be honest, is driven by fear. Many of us—and many scientists themselves—have watched parents suffer the devastations of dementia. We're frightened.

A few years ago, after I wrote a book on the teenage brain, I would sometimes give talks for juvenile justice or school groups. After a speech, I was usually driven to the airport by the person who had arranged the event. More often than not, that person, like me, was middle-aged, and as we drove along, he or she would say something along the lines of: "You know, you should write a book about *my* brain; my brain suddenly is horrible, I can't remember a thing. I forget where I'm going or why. And the names, the names are awful. It's scary."

I would smile and nod in agreement, thinking of my own middle-aged brain. Where *do* all those names go? Do they float out of our heads and into the trees? Are they up there bouncing around the interstellar clouds, gleefully watching us fumble about? And *is* this the start of something truly awful?

Not long ago, the writer Nora Ephron, who at sixty-seven was at the outer edge of what's considered the modern middle age, wrote an essay about all this called "Who Are You?"

"I know you," she wrote. "I know you well. It's true. I always have a little trouble with your name, but I do know your name. I just don't know it at this moment. We're at a big party. We've kissed hello. . . .

You've been to my house for dinner. I tried to read your last book. . . .
I am becoming desperate. It's something like Larry. Is it Larry? No it's
not. Jerry? No it's not. . . I'm losing my mind. . . ."

Originally, I shared such concerns. My aim was to find out where
the names go, the Larrys, the Jerrys, the "who are you's." From a
neuroscience point of view, I wanted to know if those names were
hidden somewhere, a brain equivalent to the secret hole in the universe where all the library cards, favorite pens, and glasses disappear.
I wanted to find out what was going wrong in middle age, and what
it meant.

After all, it's more than just memory and names. Our brains at
midlife have other issues as well. Sometimes when I'm driving now,
I look up and realize that I've not been paying the slightest attention
to the road but instead have been thinking of something else entirely,
like how I'm going to brine the turkey for Thanksgiving. The smallest interruption can be distracting, my brain flitting away from what
it was doing and off into another land. Just the other day, while
packing for a trip, I spent five frustrating minutes looking for my
toothbrush to put in my suitcase only to find that I had, just minutes
before, already *put* my toothbrush in my suitcase. After I'd packed
it, I'd gotten distracted looking for a sweater and, whoosh, all
thoughts of toothbrush-already-in-suitcase were swept out of my
head.

It would be nice to say that this kind of thing happens rarely. In
fact, it happens all the time. And while other ages have their troubles,
too—one would hardly call your average teenager a model of mind-
fulness, for instance—the changes in my brain now seem to have a
qualitative difference. In areas of memory and focus, in particular,
a tipping point has been reached—a point at which I now find myself
in a kind of automatic way relying on my twenty-something daughters not only to remind me of things I fear I'll forget but also to bring

my mind back to where it started. What *was* I talking about? At middle age, we know we're different. We know our brains are different. What has happened? Where have our minds gone? From a neuroscience perspective, are we all—bit by bit—losing our minds?

In the end, I spent considerable time tracking down the lost names, and I will tell you where they go and—according to current thought anyhow—what it all means. I also dug into the latest science on our tendency to lose our train of thought as well. Over the past few years, scientists have begun to examine this mindlessness, finding where, in fact, our middle-aged brains go when they wander off track.

Along the way, though, this book took an about-face. It's not that I forgot what I was writing about. But when I looked deeper into the latest science of the modern middle-aged brain, I found not bad news but good.

As it turns out, the brain in middle age has another story to tell that's quite the opposite of the one I'd expected. This is the middle-aged brain that we've all, in a sense, mislaid. As we bumble through our lives, it's easier to notice the bad things.

But as science has begun to home in on what exactly is happening, a new image of the middle-aged brain has emerged. And that is this: Our middle-aged brains are surprisingly competent and surprisingly talented. We're smarter, calmer, happier, and, as one scientist, herself in middle age, put it: "We just know stuff." And it's not just a matter of us piling facts into our brains as we go along. Our brains, as they reach midlife, actually begin to reorganize—and start to act and think differently.

In the end, the brain I had not expected to find was the brain I wanted to write about: this middle-aged brain, which just as it's forgetting what it had for breakfast can still go to work and run a multinational bank or school or city, a whole country even, then return home to deal with cars that talk, teenagers who don't, sub-prime mortgage meltdowns, neighbors, parents.

This is a brain—a grown-up brain—that we all take for granted. In a way, it's quite understandable. As we live longer, middle age is a moving target. A lot is not yet clear. Recently, columnist William Safire was taken to task by a reader for calling the actor Harrison Ford middle-aged at 64. "If he were literally middle-aged, then he could expect to live to 128," the reader pointed out. "By describing themselves as middle-aged, are not those in their 60s and even 70s guilty of some rather over-optimistic math?"

Most researchers locate modern middle age somewhere between the ages of forty and sixty-eight. But even that's a bit squishy. As life spans continue to stretch, what's the end and what's the middle?

As I write this, I am, at age fifty-six, decidedly middle-aged. No one, not even me at my most optimistic, would describe me as young. And no one, with the possible exception of my children, would call me old.

So middle-aged it is. But what, at the beginning of the twenty-first century, does that actually mean? And what does it mean for my brain?

This book is an attempt to answer that question.

Over the past few years, in fact, researchers have found out a great deal about the middle-aged brain. They have found that—despite some bad habits—it is at its peak in those years and stays there longer than any of us ever dared to hope. As it helps us navigate through our lives, the middle-aged brain cuts through the muddle to find solutions, knows whom and what to ignore, when to zig and when to zag. It stays cool; it adjusts. There are changes taking place that allow us to see a fuller picture of the world, even be wildly creative. In fact, the most recent science shows that serious deficits in important brain functions—ones we care most about—do not occur until our late seventies and, in many cases, far beyond.

What's more, middle age is a far more important time for our brains than anyone ever suspected. This is when paths diverge. What we do when we're on Planet Middle Age determines what the next stop, Planet Old Age, will look like. As one neuroscientist told me, at midlife, the brain is "on the cusp." What we do matters, and even what we think matters.

Over the years, we've been trained to think that the body and the brain age in tandem. Certain bodily changes are undeniable. Despite my best efforts—the regular runs, the laps at the YMCA pool, the yoga—I'm twenty pounds heavier than I ever was before. I need glasses that correct for three different distances—reading, driving, and writing on a computer. My hair, without help, is an undistinguished brownish gray, my face has deep lines. Sometimes, catching a glimpse of myself in a mirror or a window, I think, for a quick moment, that I'm really looking at my mother.

And as we watch the hair on our heads turn gray or disappear altogether, we assume that there's equivalent decay inside our heads. It's not hard to imagine our neurons turning their own shades of brownish gray, drying up, or disappearing altogether, too.

But what's actually happening turns out to be much more complicated. And researchers—from sociologists and psychologists to neuroscientists—have discovered that middle-aged brains do not necessarily act like the rest of our bodies at all.

So what do we know?

What is known of middle age now comes to us from the results of major studies just now emerging of how people actually live their lives, as well as from research from labs all over the world that are now dissecting the experience of middle age, brain cell by brain cell.

Our brains vary greatly in terms of which functions decline and which maintain their capacities, or even reach their height, in middle age and beyond. Parts of our memory—certainly the part that remembers names—wane. But at the same time, our ability to make accurate judgments about people, about jobs, about finances—about the world around us—grows stronger. Our brains build up patterns of connections, interwoven layers of knowledge that allow us to instantly recognize similarities of situations and see solutions.

And because of our generally healthy childhoods—compared with earlier generations—most cognitive declines of consequence are not occurring for those in middle age now until much later than even our parents' generation. There's also evidence that as a group we're considerably smarter than any similarly aged groups that went before us.

Much of what I've written here is quite new. Even as I wrote the book, various interpretations of some findings were still being hotly debated.

As it's come into focus and scrutiny, middle age has attracted its own rumors, fantasies, and ghosts. With the current deeper under-standing of what actually happens, however, many of those ghosts are disappearing. The midlife crisis, for instance, that currency of cocktail-party conversation, turns out on closer inspection to have little grounding in reality. The empty-nest syndrome, another staple of our expectations of middle age, is equally rare, if not imaginary.

In fact, scientists have found that moving into middle age for most is a journey into a happier time. In particularly hard or stressful moments it might not seem likely, but around middle age, we start growing happier, and the cause may be aging itself. The positive wins out over the negative in how we see the world, in part because we start to use our brains differently. There may be evolutionary reasons for this, too. A happier, calmer middle-aged human is better able to help the younger humans in his care.

Clearly, the middle-aged brain is no longer pristine. Researchers meticulously tracking the brain as it ages in humans and animals see distinct declines in the chemicals that make our brains function—the neurotransmitters, such as dopamine, that keep us alert and on the move. There's a decrease in brain branches, where neurons communicate. There's new—very new—work that has found a whole new brain state—a default mode. This is a kind of daydreaming state of quiet and continuous inner chatter where our brains increasingly go as we age, leaving us distracted, and confirmation of its existence is considered one of the most important discoveries ever made about how brains operate and age.

What's more, one scientist at Pomona College in California has now carefully documented what, in fact, happens when we forget names, why it starts in middle age, what it might mean, and why, for heaven's sake, we can remember that a person works as a banker but cannot remember that his name is Bob. There is now general agreement that some brain functions simply do not keep up, particularly what scientists like to call processing speed. If you think, at age fifty-five, that you'll be able to keep pace in all areas with an average twenty-five-year-old—to swerve as quickly to avoid a squirrel in the road or adjust as quickly to yet another new computer system at work—think again.

But in the end, a name here or there or a top rate of brain speed may not matter so much. While losses occur by middle age in our brains, they are neither as uniform nor as drastic as we feared. Indeed, even the long-held view that our brains lose millions of brain cells through the years has now been discounted. Using brain scanners and watching the brains of real people aging in real time, researchers have now shown that brain cells do not disappear in large numbers with the normal aging process. Most stick around for the long haul and, given half a chance, can be there—intact and ready—well into our eighties and nineties and perhaps beyond.

Neuroscientists at UCLA and elsewhere can now watch parts of brain cells—in particular, the fatty white coating of neurons called myelin—continue to grow late into middle age. As myelin increases, it builds connections that help us make sense of our surroundings. This growth of white matter, as one Harvard scientist has put it, may in itself be "middle-aged wisdom." There's new interest, too, in defining what exactly wisdom is. We talk glibly of someone being wise, but what does that mean? How is such a thing stored in a brain and made use of in the day-to-day life of a fifty-year-old mother of teenagers or a sixty-year-old professor? For many years, what we call experience was also taken for granted. But experience is now being broken into its component parts and we're learning exactly how experience physically changes the brain, which kinds of experience alter the brain for the better, and what it really means to be a competent manager, a prudent pilot, or a gifted teacher.

There are recent findings, too, that show how the middle-aged brain—rather than giving up and giving in—adapts. As we age, our brains power up, not down, and use more of themselves to solve problems. And it is those with the highest functioning cognitive skills who learn to use their brains this way. In some cases, as researchers at Duke University and elsewhere have found, people in middle age begin to use two sides of their brains instead of one—a trick called bilateralization. Those who recruit—or learn to recruit—the strength of their brains' powerful frontal cortex, in particular, develop what scientists call "cognitive reserve," thought to be a buffer against the effects of aging. This is the kind of brain strength that helps us get the point of an argument faster than younger peers—to get the gist, size up a situation, and act judiciously rather than rashly.

This brain reserve may also help us ward off early outward symptoms of diseases such as Alzheimer's. And there are strong hints that something as simple as education—or working—may be the key to building this brain buffer for a lifetime.

The question this leaves us with, of course, is, how can we both develop that buffer and keep it. If we're lucky enough to remain relatively healthy, can we push our brains to remain strong beyond middle age? To get that answer, science first has to tease out exactly what constitutes normal aging and what is pathology and illness. Since for years most aging research was conducted largely in nursing homes, we've had an overly negative view of what it means to get old. For many years, even most doctors thought dementia was inevitable.

But now we know that dementia, while its risks certainly increase with age, is a specific disease. If we maintain a normal path of aging without major illnesses, our brains can stay in relatively good shape.

So what do we need to do?

In the last part of the book, I explore the science of brain improvement, an area steeped in hype. What do we really know about the magic of eating blueberries or omega-3's anyhow? Does exercise make a difference, and, if so, what kind and how?

At Boston University Medical School, neuroscientist Mark Moss is studying middle-aged monkeys to find out how normal aging happens and what can keep middle-aged brains intact. Is it fish oil? Red wine? Hours on the elliptical trainer? Elsewhere scientists are testing starvation diets to see why low-calorie diets seem to prolong lives, or why poor diets, high in fat and sugar, are harmful. One top researcher at the National Institutes of Health, for instance, has been severely limiting his own caloric intake since he was in graduate school, to see if he can maintain his brain's vitality, ward off disease, extend his own life—and figure out how to prolong ours, too. Newer studies are asking what it is about obesity or high blood pressure that might increase the risks of dementia. Far beyond simply suggesting that a glass of wine or a bunch of blueberries is beneficial, researchers are now looking closely at the chemical makeup of certain foods. Is it the dark color of the fruit's skin that helps our cells stay healthy? Is it the antioxidants? How many glasses of wine do we have to drink anyhow? Can we find a pill that will work instead?

One way to measure how excited a particular group of scientists is about the potential of their field is to follow the money. And there is now real money behind various ideas about how to extend the useful life of our brain cells. Now that science knows that we do not lose millions of neurons as we age, it seems suddenly plausible that we can, if we look hard enough, find easier ways to keep our brain cells in top form. There's increasing talk of "druggable" targets to help the brain as it ages, and a number of top scientists have begun their own companies in the hopes that once that target is found, there will be money to be made. Indeed, one top researcher I know said the biggest change she's seen over the past few years has been that legitimate scientists are now talking unabashedly about possible brain "interventions," including drugs that may be within reach.

For many researchers working on the aging brain, this new culture of possibility is a surprise. But then, as we watch ourselves age, many of us, too, are finding that we have to reconsider how we think about our own brains—and our own lives—as we enter and traverse middle age.

In an essay in 2007, author Ann Patchett expressed her own surprise at the evolving talents she has found in her brain as she reaches middle age. Even as her skin droops, Patchett has discovered that her mind is maturing.

"I was searching through files of photographs recently . . . when I found the proof sheets from a photo shoot I had sat for in 1996," she wrote. "I was 32 years old, and I looked good. I mean really good: clear-eyed, sharp-jawed, generally lanky and self-possessed. . . .

"Looking at them now . . . I was struck by the fact that even though I am devoted to yoga and eat and get loads of rest and take vitamins and do all the other things you're supposed to do to maintain the lustrous beauty of youth, I looked much better 11 years ago."

But "I was also struck by the fact that I am smarter now. . . . My mind . . . is like a bank account and every investment I make seems to grow with a steady rate of interest. I am hoping that it will be there to keep me company as I age and that it will remain curious and agile. I'm working hard on it. And I do so love the work."

As I wrote this book, I, too, began to view my own brain with a new respect.

When you actually take a moment to watch what a middle-aged brain does—and does with ease—it can come as a surprise. But it is also comforting. Over and over, when I told others I was writing a book about the brain in middle age, I would be met with suspicious glances. Then, after a moment, those same people, all middle-aged, would say things like, "Well, you know I am a better teacher now," or, "Oh, well, yes, I am a better parent now." Certainly, during middle age, we have a lot going on, a lot on our minds. But many of those in middle age told me that, rather than just feeling over-whelmed, they are, on some level, quite proud of what they can accomplish. One sixty-year-old friend put it another way: "My brain feels like one of those blueberries they keep telling us to eat," she said. "You know, finally ripe and ready and whole."

And that leaves the final—and perhaps most important—question. And that is, if our brain does in fact retain its strength—

and we find methods of maintaining that strength—what shall we do with it?

The trappings and timetables of our lives are woefully out of date—set up for long-ago life spans in which by middle age we were expected to curl up—and give up. But if—as current trends indicate—many of us manage to live well into our eighties and nineties, and if we manage to keep our brains intact during that time, what will we be doing?

The world is set up to treat a middle-aged brain not as ripe, ready, and whole, but as diminished, declining, and depressed. We set up mandatory retirement ages that have little bearing on current lives. We tell teachers, lawyers, writers, and bankers they're too old to work and we send them home—to do what?

Part One: The Powers That Be

1 Am I Losing My Mind?

Sometimes, but the Gains Beat the Losses

I'm standing in my basement.

I've come downstairs to get something. The question is, what?

I look around, trying to jog my memory. I stare at the shelves where I store big pots and pans. Was it the pasta plate? My mind is suddenly, inexplicably, blank.

I stare at my hands. Maybe if I look at my hands long enough, I'll get a picture in my mind, a clue as to what I came down to the basement to put into those hands.

This is maddening.

I consider going upstairs and starting over, back to the kitchen to survey the scene to figure out what's missing, like one of those children's puzzles where, after looking at a picture, you then look at a second picture and try to find what has been removed from the first one—a tree missing a branch or a man who is no longer wearing his hat.

I don't want to go back upstairs. That's ridiculous. I stare at the shelves again. Lightbulbs?

Nothing. Nada. Zippo.

I give up and walk back upstairs. I scan the kitchen.

And then I see it—the empty paper towel holder.

Agghh!

I turn and go down the basement stairs again, this time repeating to myself over and over:

"Paper towels paper towels paper towels paper towels."

Ahh . . . the middle-aged brain. It can be bad out there.

My own most recent worst case was when I tried—really tried—to get a book for a book club I'm in. I went online and carefully ordered *The Alchemist* by Paulo Coelho. Then, a week later, I had a free moment at work and I thought, Oh, I should order that book club book. I went online and carefully typed in an order for *The Alchemist*—again.

Then a few days later, jogging in the park, a faint bell went off in my head and I thought, I think I ordered the wrong book. At home, I checked my e-mail and, sure enough, we were supposed to read *The Archivist* by Martha Cooley.

I'd ordered the wrong book—twice.

And that wasn't the end of it. Later that week, I was talking with a fellow book club member, a neurologist, who, after hearing my embarrassing story, started to laugh. It turned out that she'd gone to the library to get the book club book and had just as carefully come home with a copy of *The Alienist*, by Caleb Carr.

So there you go. Two middle-aged brains, three wrong books.

———

And that's just the beginning.

One woman I know, who is fifty-three, says she now wakes up uncertain what day it is. Another friend, also in his early fifties, finds himself dishing out guidance to his children only to be told that he had dished out the exact same advice just hours before. "They tell me, 'Dad, you told us that this morning, don't you remember?'" Well, he doesn't. And he wonders, what does it mean? Maybe he's just too busy, with the job, the kids. Maybe his children are just being annoying, playing childlike annoying tricks. Maybe—and this is not a good thought at all—he is losing his mind.

We all worry about getting old. We all worry about getting sick. But we really worry about losing our minds. Will we forget to tie our shoes or zip our flies? Will we fumble our words and fall into our soup? Are our brains on an inevitable downward slide?

It seemed, as I reached middle age—landing unprepared on the foggy planet of lost keys and misplaced thoughts—that this, sadly, was the case. But then I noticed something else. At work, at home, with friends, I was surrounded by people who knew what they were doing. These were people, also in the thick of middle age, who, despite not remembering the name of the restaurant they just ate in or the book they just read, were also structuring complex deals between oil companies on different continents and coming home to cook Coquilles St. Jacques. These were people who could simultaneously write an e-mail to a daughter who was unhappy at college, sort expenses, and participate in a conference call with colleagues in Washington.

Take Lynn, for instance. An accomplished woman in her early fifties, she has raised two children and managed a competitive and creative career for the past thirty years. There are times when she feels hopelessly muddled, forgetting where she took the dry cleaning or if she called the dentist. At other times, she told me, she feels that she "can do anything."

"I guess I'm getting older and, sure, I can tell," she told me. "But also, if I think about it, I also feel unbelievably capable." A book editor in his early fifties reported a similar mixed sense. "You know," he said to me at lunch recently, "when my daughter started taking piano lessons, I decided to take lessons with her. Boy, it's hard to see her learn it so much faster than I can. I sit there and watch and I think, What happened to my brain?

"But it's weird," he went on. "I have to say that I also feel much smarter these days. I know what I'm doing at work. Nothing seems to faze me. I feel truly competent."

Not long ago, when I told one of the editors at my newspaper that I planned to write about the middle-aged brain, he laughed, thinking of his own fifty-eight-year-old talents. "Oh, my," he said. "The middle-aged brain. That's really interesting because sometimes it really seems like there's not much left up there. You know, the synapses are not synapping like they used to."

Then, when I looked down at his desk, there was this complicated chart, full of boxes and arrows and circles. His middle-aged brain, with its unsynappy synapses, had taken on what was then the most complex issue the company had faced—how to integrate the new Web operations with the old print infrastructure. He took on this task amid his other duties, such as finding money for the paper's continued coverage in Baghdad. Undaunted, he was handling this thorny job with, as they say in Spain, his left hand. What's more, he mentioned by way of passing conversation that he'd just helped plan the weddings of two of his daughters, one in the Midwest, hardly a task for a brain on the brink of extinction, I thought.

A short time later, I was having dinner with another friend of mine, Connie, now in her early sixties and working as an editor. She, too, has a full-tilt life—a daughter in college, a mother who recently died after a long illness, a book under way, and recent bouts in her family with two life-threatening diseases. As we drank our red wine, we spoke about how our own middle-aged brains were doing. She had her concerns. She pulled her hands in front of her face like a curtain closing to illustrate what sometimes happens to her now when "whole episodes" of her day seem to vanish from her brain cells. At times, too, she has to stop herself as she starts to put the bananas in the laundry chute. Still, when I asked her if she also feels more with it in other areas, her face lit up.

"I guess I take it for granted," she answered. "Sometimes now I just seem to see solutions. They pop into my head. It's crazy. Sometimes, like magic, I am brilliant."

Consider, too, Frank. At fifty-five, Frank has come up with a little game to help his brain. When he can't recall the names of those he just met or has known for years, a situation that happens with greater frequency, he rapidly runs through the alphabet, trying to match a letter to a name to jog his memory. "You know, *A*, is it Adam? No, *B*, Bob. Yes, that's him, Bob Smith. That's what I do," he said. While he is priming his brain with tricks, Frank also finds that in other, far more important ways his brain is functioning better than ever. As the chief financial officer for a nonprofit organization in New York, he spends his days wrestling with one knotty management tangle after another. And over the years, he finds these challenges getting easier, not harder. Often he sits with another manager and they'll toss ideas back and forth about how to size up and solve a problem.

Both have been managers for years and, with all those years of experience etched in their brains, they speak a kind of shorthand, saying, "Hey, you know he is the type that . . . and you know we really ought to move that over there. . . ." They can often finish each other's sentences, in a language that Frank says younger people with less experience in their brains simply would not get.

"We understand each other, but more important, when we talk we get somewhere. We actually solve problems. When situations come up now, I have a whole library of experience to draw on to figure out what to do. . . . I guess you would call it, what, expertise?"

Science Changes Its Mind

Indeed, while the buoyancy of the middle-aged brain may be a surprise to many of us, it's no longer a surprise to science. After years of believing that the brain simply begins to fade as it ages, a more nuanced picture has begun to emerge. While many of us would simply chalk up Frank's experience to experience and leave it at that, neuroscientists—perhaps the most skeptical crowd around—have found that the brain at middle age has its own identity and surprising talents. Experience—and expertise—has literally changed our brains.

By middle age, the brain has developed powerful systems that cut through the intricacies of complex problems to find, as Frank does, concrete answers. It more calmly manages emotions and information. It is more nimble, more flexible, even cheerier. Equipped with brain scanners that can peer into brains as they age, neuroscientists find executive talent and, even more encouraging, what they call cognitive expertise.

Analyzing long-term studies of actual people as they have aged, psychologists are now realizing that our long-held picture of middle age has been incomplete and misleading. One new series of fascinating studies suggests that it may be the very nature of how our brains age that gives us a broader perspective on the world, a capacity to see patterns, connect the dots, even be more creative.

Certainly, there are times when the patterns we see are missing a few pieces.

One recent morning, I found myself yelling (politely) at my husband, Richard.

"I thought you were going to buy milk," I said, as I looked in the refrigerator while he was in the bedroom getting dressed.

"I did," he said.

"But it's not here," I said, staring at refrigerator shelves that were, indeed, milk-less.

This brought Richard to the kitchen to see for himself.

"But it's right there," he said, pointing to the milk carton sitting on a counter behind me. "You just put some in your cereal."

Sure enough. I had, in fact, gone to the refrigerator, gotten the carton of milk, and poured some on my cereal. Then, after busying myself with another activity—making tea—I'd become distracted and the image of the milk on the counter had disappeared from my brain.

And such difficulties are not imaginary. The brain at middle age is not protected from harm. We develop schemes like Frank's game for figuring out what a person's name is because, in fact, we have more difficulty with name retrieval, particularly the names of those we've not seen in a while. Connections that tie faces to names weaken with age. Our brains slow down a bit, too. For instance, if chess players compete in a game that depends on speed—say, they're given a few seconds to move a piece—younger players always beat older players. In brain-scanning studies, scientists can watch the middle-aged brain as it loses focus and begins to wander aimlessly.

For many years, a major line of thinking was that the brain becomes more easily distracted with age simply because age brings so many distractions. Even now, I hear this explanation from some who insist that their brains may miss a beat now and then simply because their circuits are overloaded.

"I hate it when people say they are having a senior moment," said one woman I know in her early sixties. "People lose their keys when they are my age and they think it's their aging brain. But plenty of teenagers lose their keys, and when they do, they just, well, they just say they lost their keys."

Such explanations are enticing, and have some truth to them. By middle age, we ask a lot of our neurons—we relearn geometry to help our teenagers with their homework, we find ourselves as out-patient hospital managers as our parents fall ill, we untangle competing egos and agendas at work, we decipher unintelligible fine print in refinancing applications—all pretty much at the same time that we begin to *really* worry about a whole host of events on an even larger scale: Will polar bears completely disappear with global warming? Will Pakistan use a nuclear bomb? Should we negotiate with Iran?

Until recently even many of the scientists thought information overload *was* the problem. We have a lot to do, and we simply get overtaxed and overwhelmed. With all we have careening around in our neurons, no wonder we lose our focus.

But such explanations are no longer considered sufficient. Over the past few years, science has taken a more serious look at our middle-aged brains and found that, in some areas, declines are real.

In truth, we know that, too. A friend who is fifty-five said she battles her brain every day now.

"I used to be able to keep a mental note of everything. I was really organized and I just had it all in my head, what I had to do for work, with my boys," she told me. "Now I have to write everything down and I still get confused. I keep looking for my glasses when they're on my head—that kind of thing happens all day long. Sometimes I just feel like my brain is fried."

By middle age, we all have similar stories—and worries. But the latest science is reassuring. It's true that the first changes from degenerative brain diseases such as Alzheimer's often begin much earlier than we thought. But researchers have now begun to sort out the differences between the stirrings of dementia and the normal aging process. And most of us, while beset with a normal level of middle-aged muddle, are, in fact, quite normal.

What's more, we're quite smart. And, on some level—if we think about it—we know that, too. For instance, my friend who complained about battling her brain every day was recently promoted to a new, high-level job that involves intense scrutiny of detail. And despite her middle-aged brain—perhaps *because* of her middle-aged brain—she's already handling that job with ease. She knows what to pay attention to and what to ignore. She knows how to get from point A to point B. She knows what she's doing.

———

The middle-aged brain is a contradiction. Some parts run better than others. But perhaps more than at any other age, our brains in middle age are more than the sum of their parts.

In fact, as we shall see, long-term studies now provide evidence that, despite a misstep now and then, our cognitive abilities continue to grow. For the first time, researchers are pulling apart such qualities as judgment and wisdom and finding out how and why they develop. Neuroscientists are pinpointing how our neurons—and even the genes that govern them—adapt and even improve with age. "I'd have to say from what we know now," says Laura Carstensen, director of the Stanford Center on Longevity at Stanford University and a leader of the new research, "that the middle-aged brain is downright formidable."

A friend who is a poet told me recently that she does not think that she could have written the poetry she does until she had reached her mid-fifties—until her brain had reached its formidable age.

"It feels like all the pieces needed to come together," she said. "It's only now that my brain feels ready. It can see how the world fits together—and make poetry out of it."

2   The Best Brains of Our Lives

A Bit Slower, but So Much Better

Here's a short quiz. Look at the following list:

January February March April January February March May January February March June January February March—

What would the next word be?

Got it? Now, how about this one:

January February Wednesday March April Wednesday May June Wednesday July August Wednesday—

What would the next word be?

Now try it with numbers. Look at this series:

1 4 3 2 5 4 3 6 5

What would the next number be?

Did you get them all?

These are examples of questions that measure basic logic and reasoning. The answers are, in order, July, September, and, for the number sequence the next number would be 4 (and then 76. The series goes like this: 1-43 2-54 3-65 4-76 and so on).

Such problems test our abilities to recognize patterns and are routinely used by scientists to see how our cognitive—or thinking—processes are holding up. And if you're middle-aged and have figured out all of them, you can be proud—your brain is humming along just fine.

Indeed, despite long-held beliefs to the contrary, there's mounting evidence that at middle age we may be smarter than we were in our twenties.

How can that be? How can we possibly be smarter *and* be putting the bananas in the laundry basket? Smarter and still unable, once we get to the hardware store, to remember why we went there in the first place? Smarter and, despite our best efforts to concentrate on one thing at a time, finding our brains bouncing about like billiard balls?

To begin to understand how that might be, there is no better person to start with than Sherry Willis. A psychologist at Pennsylvania State University, Willis and her husband, K. Warner Schaie, run one of the longest, largest, and most respected life-span studies, the Seattle Longitudinal Study, which was started in 1956 and has systematically tracked the mental prowess of six thousand people for more than forty years. The study's participants, chosen at random from a large health-maintenance organization in Seattle, are all healthy adults, evenly divided between men and women with varying occupations and between the ages of twenty and ninety. Every seven years, the Penn State team retests participants to find out how they are doing.

What's important about this study is that it's longitudinal, which means it studies the *same* people over time. For many years, researchers had information from only cross-sectional human life-span studies, which track different people across time looking for patterns. Most longitudinal studies, considered the gold standard for any scientific analysis, were not begun until the 1950s and are only now yielding solid information. And they show that we've been wildly misguided about our brains.

For instance, the first big results from the Seattle study, released just a few years ago, found that study participants functioned better on cognitive tests in middle age, on average, than they did at any other time they were tested.

The abilities that Willis and her colleagues measure include vocabulary—how many words you can recognize and find synonyms for; verbal memory—how many words you can remember; number ability—how quickly you can do multiplication, division, subtraction, and addition; spatial orientation—how well you can tell what an object would look like rotated 180 degrees; perceptual speed—how fast you can push a button when you see a green arrow; and inductive reasoning—how well you can solve logical problems similar to those mentioned above. While not perfect, the tests are a fair indicator of how well we do in certain everyday tasks, from deciphering an insurance form to planning a wedding.

And what the researchers found is astounding. During the span of time that constitutes the modern middle age—roughly age forty through the sixties—the people in the study did better on tests of the most important and complex cognitive skills than the same group of people had when they were in their twenties. In four out of six of the categories tested—vocabulary, verbal memory, spatial orientation, and, perhaps most heartening of all, inductive reasoning—people performed best, on average, between the ages of forty to sixty-five.

"The highest level of functioning in four of the six mental abilities considered occurs in midlife," Willis reports in her book *Life in the Middle*, "for both men and women, peak performance . . . is reached in middle age.

"Contrary to stereotypical views of intelligence and the naïve theories of many educated laypersons, young adulthood is not the developmental period of peak cognitive functioning for many of the higher order cognitive abilities. For four of the six abilities studied, middle-aged individuals are functioning at a higher level than they did at age 25."

When I first learned of this, I was surprised. After researching the science on the adolescent brain, I knew that our brains continue to change and improve up to age twenty-five. Many scientists left it at that, believing that while our brains underwent large-scale renovations through our teens, that was about it. I, too, thought that as the brain entered middle age, it was solidified and staid, at best—and, more likely, if it was changing in any big way, was headed downhill.

After speaking with Willis one afternoon, I went out to dinner with friends and couldn't resist talking about what was still whirring in my head. "Did you know," I asked the middle-aged group over pasta and wine, "that our brains are better—*better*—than they were in our twenties?"

The reaction was swift.

"You're crazy," said one of my dinner companions, Bill, fifty-two, a civil engineer who owns his own consulting firm. "That's simply not true. My brain is simply *not* as good as it was in my twenties, not even close. It's not as fast; it's harder to solve really hard problems. Come on, if I tried to go to Stanford engineering school today, I would be toast. *TOAST!*"

Bill is not wrong. Our brains do slow down by certain measures. We can be more easily distracted and, at times, find it more taxing to tackle difficult new problems, not to mention our inability to remember why we went down to the basement.

Bill does not have to go to school anymore, but even in his day-to-day work he compares his current brain to his younger brain and sees only its shortcomings. However, Bill is *not* seeing that his brain is far more talented than he gives it credit for. If you look at the data from the Willis research, the scores for those four crucial areas—logic, vocabulary, verbal memory, and spatial skills—are on a higher plane in middle age than the scores for the same skills ever were when those in her study were in their twenties. (There are also some interesting gender gaps. Top performance was reached a bit earlier on average for men, who peaked in their late fifties. Men also tended to hold on to processing speed a bit longer and do better overall with spatial tests. Women, on the other hand, consistently did better than men on verbal memory and vocabulary and their scores kept climbing later into their sixties.)

Equating Age with Loss

So why don't we all know that? Why is Bill, along with so many of us in middle age, swallowed by the sense that, brain-wise, we are simply less than we were? In part, it's the steady drumbeat of our culture, determined to portray aging as simply one loss after another. In part, it's because for years people in aging science studied only those in nursing homes, hardly the center of high-powered inductive reasoning. Researchers simply skipped the middle.

But our own brains are not helping, either. Brains are set up to detect differences, spot the anomaly, find the snag in the carpet, the snake in the grass. So we notice changes in our own brains, too. But the differences we register in all likelihood refer to our brains of a few years ago, not the brains we had twenty-five years earlier. And when we notice slight shifts, which is certainly possible, we're convinced that our brains have been in a downward trajectory since graduate school.

In other words, we pick up on the tiny defects in the carpet but fail to notice the more subtle, gradual process that over the years has painstakingly built our brains into a high-functioning, formidable force—a renovated room.

In the Seattle study, those between the ages of fifty-three and sixty, although still at a higher level than when they were in their twenties, nevertheless had "some modest declines" compared with a previous seven-year period. This difference in certain mental abilities from the earlier years, however slight, is what we notice. But it's an illusion.

"The middle-aged individual's perception of his or her intellectual functioning may be more pessimistic than the longitudinal data would suggest," says Willis. "Comparisons . . . may be more likely to be made over shorter intervals. One may have a more vivid or accurate perception of oneself seven years ago than twenty years ago."

In short, Bill was most likely thinking of his brain being slightly worse in some small ways at fifty-two than it was at forty-five—not twenty-five—when he assessed how poorly he thought he was doing now. The result is that, like most of us, he is keenly aware of flaws and completely unaware of the overall high level of ability of his own middle-aged brain.

"Your friend Bill does not realize how well he is doing because he is a fish in water" and can't see how nice the water is, says Neil Charness, a psychologist at Florida State University and an expert in this area of research.

"Smarter and Smarter" by Generation

Of course, Bill is not the only fish in that particular body of water.

"For a long time we all thought that the peak was in young adulthood," Willis told me. "We thought that the physical and the cognitive went in parallel and, partly for that reason, we funnel our educational resources into young adulthood thinking that is when we can most profit from it. But remember all this is new. We have never had this long middle age when we are doing so much. And we are finding out new things about this new period of life all the time."

Indeed, some of the more recent research has started to break up aging into more distinct segments for examination. It is no longer just young versus old. Now we are looking even more closely at the middle years, even breaking those years into smaller segments to see how our current brains compare to those in previous decades.

A study by Elizabeth Zelinski at the University of Southern California, for instance, compared those who were seventy-four now with those who were that age sixteen years ago. She found that the current crop did far better on a whole range of mental tests. In fact, their scores were closer to those of someone fifteen years younger in earlier testing, findings that, as Zelinski points out, have "very interesting implications for the future, especially in terms of employment."

There's also a heartening downward trend now showing up in broad measures of cognitive impairment in individuals, the kind of mild forgetting that can plague brains as they age. A recent study by University of Michigan researchers found that the prevalence of this minor type of impairment in those seventy and older went down 3.5 percentage points between 1993 and 2002—that is, from 12.2 percent to 8.7 percent.

Nevertheless it's easy to be concerned. Many of us have watched parents who, instead of dying quickly by falling off cliffs or tractors, spent years dealing with the debilitating effects of chronic ailments such as heart disease or Alzheimer's.

"We have a lot of firsthand experience caring for our parents and we know we share genes with them and we watched what happened to them and we are very worried," Willis says.

When I spoke with Willis, she was on sabbatical, trying to learn a new way of analyzing human life-span data with a dizzying array of complex equations. She readily admitted to some frustration with her own middle-aged brain.

"Look," she told me, "I am fifty-nine and I have to make lists of the things I have to remember. I have to write down that I am going to talk to you and where I am going next, and now I'm trying to learn this new methodology and maybe it takes a little longer than it used to and it can be frustrating."

But she adds quickly, "I am quite proud that I am beginning to understand it and, remember, when students learn these new things they are *just* studying and nothing else. They have a whole semester to devote to it. But here I am trying to learn it and at the same time I am very, very busy. I'm answering a gazillion e-mails and shopping and writing and talking to you.

"So really, I have to tell myself, give yourself a break. There is no question, the brain does get better at middle age."

Extending her research, Willis is now digging even deeper into the folds of the middle-aged brain. Using new imaging technology, she is looking to see what kinds of structural changes occur in brain volume in middle age and if those changes affect cognitive abilities as people age. She's also trying to find out what effect such chronic diseases as diabetes and cardiovascular problems in midlife have on a person's ability to maintain high levels of brain function later on. All in all, she fully expects to find that the brains of her grown-up subjects do not stand still.

"If we are lucky," she says, "our brains continue to develop and improve."

So, if that's so, how do we do both? How can a brain at age fifty-two be wandering around the living room trying to remember what it is looking for *and* galloping along on a higher plateau than it did in college? Can we break apart that inherent contradiction further? And if so, what do we call the good aspects of our brains? Is it knowledge? Is it expertise? Is it experience? Maybe it has more to do with intuition. Or how about simple survival instincts?

More important—aside from strict cognitive tests—is it possible to measure all this in the real world?

A few years ago, the answer would have been no. But that has changed, too. Researchers have now gone looking for this middle-aged stuff—this middle-aged je ne sais quoi—and they've found it both in the real world, by following real people through their entire lives, and, increasingly, by using new scanning technology deep inside the complex structure of our brains.

One of those who have looked the hardest is Art Kramer, a psychologist and neuroscientist at the University of Illinois. A couple of years ago, Kramer decided to see if he could find out how a middle-aged brain actually functioned in day-to-day life. In particular, he and his colleagues wanted to see how a middle-aged brain would do in a job that calls for rapid-fire decision making. So they decided to look at air-traffic controllers.

In this country, air-traffic controllers must retire at age fifty-five. Many other countries let controllers work much longer and don't have more accidents than we do. Who is right? Are we somehow safer here because we insist that those in such jobs, on whose top-level brain function we rely for our safety, have a mandatory age cutoff, regardless of health or ability? Or, asked the opposite way, is it possible that by forcing retirement at age fifty-five, we're losing out on the best brains—grown-up brains—that could keep us even safer?

To test this, Kramer went to Canada, where controllers can work until they're sixty-five. There, they put a group of young and older controllers through a seven-hour battery of cognitive tests and then had them, for a long stretch of time, do work that simulated their daily jobs.

"In real life controllers work at computers, and in our simulation we used computers and we had them do all sorts of things, just as if they were working," Kramer explained when I spoke with him. "Sometimes they were really busy and talking to pilots and watching a screen and having aircraft coming in at different speeds. We also had them sequencing flight patterns. There were a lot of things for them to deal with."

And what did they find? Older controllers did just as well as their younger colleagues. "They clearly performed as well on simulated tasks as the younger group. There was no difference in level," said Kramer.

On the cognitive tests, there were differences, but they, too, were instructive. In areas such as processing speed, younger controllers did better. But in two important cognitive areas—visual orientation (the capacity to look at a plane in two dimensions on a computer screen and imagine it in three dimensions in the sky) and dealing with ambiguity (coping well with conflicting information, computer crashes, or even the possibility that the computer might be wrong)—older controllers, again, did just as well.

Studies of pilots find the same thing. In research led by Joy L. Taylor of Stanford/VA Aging Clinical Research Center and published in the journal *Neurology*, 118 pilots aged forty to sixty-nine were tested over a three-year period in flight simulators that involved piloting a single-engine aircraft over flat terrain near mountains. Taylor found that older pilots did not do as well the first time they used the simulators, which tested skills in communicating with air-traffic controllers, avoiding traffic, keeping track of cockpit instruments, and landing. But as the tests were repeated, the older pilots were actually better than younger pilots in the underlying point of the whole exercise—avoiding traffic and collisions. In other words, the older pilots took longer to catch on to the new test at first, but they outperformed younger pilots when it came to doing what was most important—keeping the planes where they were supposed to be.

"The thing is, if you have many years of experience, that serves you well and is very, very useful," Kramer says. "And if an older person maintains the skills he needs, perhaps he can perform in professions that we thought he could not in the past."

Where Expertise Finds a Home

In an odd way, of course, we think we know this, too. We talk a lot about experience, often in glowing terms. We praise it in an architect or a lawyer; we look for it in presidential candidates.

But even as we give experience its due, strangely, we overlook its true nature and impact.

Granted, this is elusive. Can you plot on a graph how well a person manages a staff? Can you count the number of times a person sagely decides to hold her tongue or, through well-practiced tact, leads a bickering group to consensus? For that matter, how do you nail down the exact moment when a parent is being an expert parent, determining whether to hug or scold a difficult child? Can you find, with cognitive tests, the enthusiasm, judgment, and patience an experienced teacher brings to his class?

It's easy to throw experience around as a catchall—and leave it at that. But that has led to an astounding lack of appreciation for the very place where such experience makes its home—in middle-aged brains. All those years of know-how and practice and right-on-the-money gut feelings aren't, as one researcher put it, "building up in our knees."

Over the past few years, there has been an attempt to address this neglect. A whole field has developed to pin down what scientists like to call "expertise." This does not completely capture the whole nature of what we call experience, either. But it certainly takes some steps in that direction.

Neil Charness, for one, has spent his career looking at all this. Now fifty-nine, Charness first got interested in what makes aging brains retain their power at his first job when he studied bridge players.

Although the prevailing view had been that older bridge players were slower and had poorer memories and were, therefore, weaker players, Charness found, in a sample of real people playing real games, that simply wasn't true. He found that if the task in the game required mostly speed, the older players performed at a lower level than younger ones. But in the most fundamental task in bridge—basic problem solving—older players "could easily play at high levels."

Some argue that brain-processing speed is so fundamental to the brain that a decline that comes with age fouls up the works overall, making all functions worse, but Charness and many other neuroscientists are now convinced otherwise.

"So we were left with a kind of paradox," Charness explained when I spoke with him. "We had tended to think that one skill—processing speed—underlies all skills, but this study helped raise my awareness that that was not true."

Most recent research in this area has focused on bridge and chess because their outcomes are easy to measure. And Charness says research continues to show that while age takes its toll on the speed of older players, that specific decline in our brains, which begins in our twenties, does not affect overall performance.

"There's no question that players slow down, but if what you are doing depends on knowledge, then you're going to do very well as you get older," Charness says. On average, it takes ten years to acquire a high level of skill in a whole range of areas, from golf to chess. "And it makes sense," he says. "Which would you rather have on your team, a highly experienced fifty-five-year-old chess master or a twenty-five-year-old novice?"

There have been recent attempts to measure this talent in other real-life settings as well. And those studies, too, find that despite loss of speed and the fact that it can sometimes take older individuals longer to learn certain new skills, they navigate their work lives with increasing ease and dexterity.

One recent study found that older bank managers showed normal age-related decline on cognitive tests, but their degree of professional success depended almost entirely on other types of abilities, the kind that Charness refers to as the "acquired practical knowledge about the business culture and interpersonal relations that made a manager work more effectively." Over the past few years scientists have developed new ways to measure success in the real world by looking at what they call practical, or tacit, knowledge. One way they do that is to give managers actual scenarios followed by different solutions that have been shown to work or not work in professional settings. Once study participants choose their solutions their scores are rated. And in this case, as in many other similar studies, older workers, calling on their richly connected, calm, pattern-recognizing middle-aged brains, consistently won expert ratings.

And in some ways, our brains are increasingly being given a cultural boost as well. It is not just biology that's helping. For many years, many people thought that midlife brought only depression or declines in energy or zeal. But now we know that such difficulties can—and do—occur at all ages, not just in the middle years. As the average life span has lengthened, we now have plenty of people growing older in fine cognitive and physical shape whom we can not only look to as role models but also study to figure out what actually takes place to make that happen. While some parts of all this—including, in some cases, our own perceptions of ourselves, as well as the official world of employment—have lagged behind, overall attitudes show signs of a shift. There are more people who are simply not giving in or giving up. And science, increasingly, backs them up.

There is now, for instance, a growing field of study that seeks to figure out how, precisely, to maintain peak performance as we age.

It used to be assumed that high levels of achievement at any time of life was mostly a result of luck and genes, with effort only a small part of it all. But it turns out that continued success has much less to do with inborn genius and more to do with what Charness and his colleagues now call "deliberate practice," a commitment to working at a skill over and over and meticulously zeroing in on faults—the kind of strategic practice that can work at any age.

And science is also now showing how as we age we develop compensatory tricks when necessary. Many of the best baseball pitchers start their careers as fastballers, relying on lightning speed to work their magic. But as time passes, and the edge comes off those ninety-eight-mile-per-hour throws, they adapt and fully develop other pitches—curves, sliders, breaking balls—to remain competitive. The fastball is still there, it's just not as fast—and the most talented use their wiles to remain the best. It is much the same with the middle-aged brain.

Even the pianist Arthur Rubinstein adopted new tricks as he aged. He sometimes made up for an age-related decline in movement speed by slowing down *before* a difficult passage to, as Charness says, "create a more impressive contrast."

And the good news is that such masterful skills, for the most part, accumulate naturally, especially in our multilayered modern world. The simple act of survival—in the course of living and making a living in our challenging environment—may make our heads ache, but it also strengthens what's inside our heads.

As Sherry Willis says: "I think the scores are so much higher in midlife than in young adulthood because we have had so much more life experience, especially on the job. Even computers are helping us to be more logical linear thinkers. The job environment is an intense learning environment, much more intense than when we were in school.

"And it's odd to think that the brain would *not* continue to develop," she adds. "Most professional jobs are very stimulating and complex and, even in leisure time, we have more opportunities to take up complicated things like photography. All that complexity can bring on what we call stress, of course, but I think that if we can handle that emotionally, it might all be very good for us—and good for our brains."

If Considered—Appreciated

And while outright appreciation is rare, many of those in middle age, when pressed, do offer a surprisingly glowing testament to their own brains.

Brad Burtner, an air-traffic controller, told me that as far as he's concerned, there's "no question" that he's a better air-traffic controller now than he was when he was younger. "Definitely, definitely, I would say I am much, much better at the job now," he said.

After nearly thirty years as an air-traffic controller, most of them spent working at the large international airport outside Cincinnati, Burtner was forced to retire in 2008 after he reached age fifty-five. A marathon runner who jogs twelve miles a day, Burtner considers the idea of retiring in middle age so silly he plans on continuing his work at a nearby small private airport that does not have age cutoffs.

This is not to say that he, like most of us, doesn't notice some missteps. When I spoke with Burtner, he had no trouble ticking off his brain's deficits. "I clearly have more problems with my memory now," he said. "When I was younger I could keep all the different altitudes of all the planes in my head, thirty planes at a time. And now I can't do that. I have to write them down.

"But, you know," he added, "that's the way we are supposed to do it anyhow and it's probably safer. If I had a stroke someone could come up and see where every plane was because I am so careful about writing it down now."

Burtner sometimes finds it harder to concentrate. "I think I am more easily distracted than I used to be," he said. "But I know that and I make changes. If someone has the radio on, I will say, hey, could you turn that down?"

And even with those concerns, Burtner insists he is a far safer controller in his fifties than he was in his twenties or even his early thirties.

As Neil Charness says: "The simple fact that older workers do just as well as younger ones in overall performance, despite fairly predictable declines in speed, is a testament to how important these other abilities are."

Like my friend the poet, Burtner finds he can do his work better largely because it is only now that "all the pieces come together."

"Now I anticipate situations before they happen. And I always have a backup plan. If there's a thunderstorm, I know what I'm going to do if the first plan doesn't work," he told me.

"The big point," he said, "is that now I control the situation instead of letting the situation control me. Now I think about the whole situation, how things fit."

3 A Brighter Place

I'm So Glad I'm Not Young Anymore

The Santa Cruz campus of the University of California sits at the crest of a hill, a small cluster of buildings tucked into a forest of redwoods. To get there, you turn off Highway 1, south of San Francisco, leave the Pacific Ocean behind, and head up a mile-long road that winds its way up to the campus.

In the early 1970s, even those of us at nearby Berkeley considered Santa Cruz the most laid-back place of all. Of course, much of that has changed. As Silicon Valley money flowed over the mountains, roads became clogged with cars that these days are more likely to be BMWs than VW vans.

Still, on the bright warm February day that I visited, I was relieved to find that Santa Cruz had not lost all of its flower-child flavor. As we drove up the hill, we passed long-haired students pedaling clunky-tired bikes, still in tie-dyed shirts. There was a sign that said, simply, PEACE CORPS, and on campus, professors sat on the ground, speaking with circles of smiling students.

Santa Cruz had aged, but in a calm and happy sort of way.

So perhaps it is fitting that it was at Santa Cruz—amid the serene and ancient trees—that a quiet effort had been under way to figure out why humans, as we age, also get happier. Indeed, scientists are finding that, starting around middle age, we begin to adopt a rosier worldview.

This was, of course, not supposed to be. Many of us grew up dreading middle age. We read John Updike's chronicle of poor Rabbit's descent into disappointment as he reached middle age, "his prime is soft, somehow pale and sour. . . . [his] thick waist and cautious stoop . . . clues to weakness, a weakness verging on anonymity" We shuddered at Gail Sheehy's message in *Passages* warning us to beware the impending doom, the "Forlorn Forties."

What happened to all that?

Well, it turns out that what actually happens is that our moods get not worse but better. In fact, our brains may be set up to make us more optimistic as we age.

It is, even now, a revolutionary view. And it was to hear about this view that I went to Santa Cruz on that day to see Mara Mather*. A cognitive psychologist, Mather is slender, short, athletic, and glowing. She has blond curly hair, light blue eyes, pale and pretty. When I caught up with her, she was sitting in her plant-filled office, sunlight streaming in through a large window. She wore gray pants, a black turtleneck, and dangling silver earrings, and if I hadn't known that she already had tucked securely under her belt hefty degrees from both Princeton and Stanford as well as a file drawer full of solid science, I would have guessed she was about fifteen years old.

Rather, when we first met, she was thirty-four, and, perhaps because she was only thirty-four, she appeared never to have been exposed to any gloomy assessment of midlife. "I don't know, maybe I was lucky," she said. "I ended up with a good view of getting older. I knew my grandmother and she was fun, vital, sociable, extroverted. When I was growing up in Princeton my great-grandfather came to visit us. He was one hundred and he was fine. I thought that's what getting old was.

---

* Mara Mather is now a professor at the University of Southern California.

"It is a bit surprising. I mean, in middle age, there's a lot of loss, I know," she said. Friends die. Parents get sick. So it's hard to think about our moods improving, but they do."

I must say, this idea seemed more than odd to me at first. In the thick of middle age myself, *cheeriness* is not the first word that comes to my mind. Stressed-out might be a better description. Most others I spoke with also greeted the idea with hefty skepticism as well. But—slowly and consistently—an alternate thought emerged.

Not long ago, for instance, I was walking to get coffee with my colleague Erica, then fifty-two. We were talking about being in our twenties, as her niece and my daughters were at that point. We talked about how incredibly hard that age is, with its ups and downs, with boyfriends in and out, the "who-am-I's" and "what-am-I-doing's."

"I would never, ever want to be twenty again; it was awful being twenty, awful," Erica said as we crossed a street in Manhattan. "Now, I know, I'm older and there's loss."

We walked a bit farther in silence. "But you know," she added after a bit, "when I think about it, it's strange, but even with all that, I've never been happier. Isn't that weird?"

Another woman, who is in her late fifties and a writer at a large magazine, told me she has never had so much to do, with a testy teenager and a mother exhibiting the early signs of dementia. But she said that she, too, noticed something new recently. For whatever reason, she now finds herself focusing less on the downside of life. "I see them, the bad things around or in my day or with my mom, but I am not quite as beaten down by them," she said.

De-Accentuate the Negative

So how can we explain this newfound serenity? Are we just so fed up with bad things that we simply shut them out? Certainly, such contentment does not at all match the picture we've been presented about how this would all play out. Where are the midlife crises? The empty nests? What is going on here?

To understand what might be happening, the best place to start is—again—inside the brain. In particular, we have to look at a tiny sliver deep in the brain called the amygdala. Even if you know nothing about the brain—or think you know nothing—you are nevertheless quite aware of your amygdala. This is your body's Homeland Security Department. If you see a scary-looking fellow plane passenger, have to talk with your boss about your performance, even speak with your teenager about sex, it's your amygdala that goes into action, revving up the rest of the body to make that crucial call: fight or flight?

The amygdala is a primitive part of the brain. It is small. (Well, technically that should read, "They are small." You have two, one on either side of your brain, and in proper plural they are called amygdalae, or "almonds" in Latin, named for their shape and size.)

So what could this ancient alert system, set up to keep early humans away from rampaging lions, be up to in our modern middle age? Not long ago, Mara Mather set out to find out. Working with Laura Carstensen, the Stanford psychologist, and neuroscientist John Gabrieli, now head of a brain-imaging lab at MIT, Mather and her colleagues found—after scanning the amygdalae of young and old—that as we get older, in a remarkably linear fashion, we, and our amygdalae, actually react *less* to negative things.

Over and over, Mather and the other researchers tried to get older people to take the negative view. While they lay in brain scanners, those in the study were shown pictures of standard scenes that are known to elicit positive reactions—puppies, children on the beach— and scenes that trigger negative responses—cockroaches crawling on pizza, people standing over a grave.

And over and over, the positive won out. As we get older, our amygdalae respond less and less to negative stimuli. And since the amygdala is pretty much set up to respond the *most* to the negative, this finding is extraordinary. Even Gabrieli says he was taken aback by the strength of the results, which were compelling. And it's important to remember that the brain scans were detecting changes in the amygdalae long *before* the people ever became conscious of what they were seeing. Indeed, those in the study had no idea what their brains were doing.

"We are seeing the moment of perception," Gabrieli says. The study found that our brains—in some automatic, preconscious way—begin to, as they say, accentuate the positive and eliminate the negative.

To see how impressive this is, it helps to know a little context. For years it was simply assumed that as we aged—and our bodies started to slow—our emotions would generally follow suit, all becoming fainter as the years went by. On one level that view held sway with scientists because it seemed to make perfect sense.

But, like a lot of what we thought we knew about the brain, that, too, was wrong. Indeed, when the study of aging began in earnest (the serious study of aging is only a few decades old), quite the opposite turned out to be the case. As we age, our emotions not only remain largely intact but are also considerably more robust than our abilities in some other areas, such as how well we recall certain facts. As we get older, for instance, it's easier to pinpoint how we felt on a given day—"I felt sad"—than what was actually happening—"it was raining."

But even that research still got one large part of the picture upside down. It assumed that if our moods stayed strong, the strongest moods would be the negative ones. Early aging researchers, as we've said, based nearly all their work in nursing homes and, not surprisingly, found considerable grouchiness.

Luckily, as she started her own investigations, Mather didn't even think of looking in such places. She had a more open mind. And when she arrived at Stanford to do postdoctoral work, she found that Laura Carstensen not only had a mind as open as hers but was already deeply engaged in upending long-held views of how our brains act as we age.

"I got to Stanford and Laura was doing all these incredible studies about aging and I was interested in memory and it just seemed natural," Mather said.

At first meeting, Carstensen, too, hardly seems a scientific revolutionary. When we first met for lunch at the elegant faculty dining room at Stanford University, she looked—with a swath of white hair at her forehead—every bit the serious university professor that she is. But this was not the full picture. As I got to know Carstensen and began to appreciate her instincts for looking at issues in new and different ways, I began to think of her as a kind of Che Guevara of science, determined to, as she says, "change the nature of aging."

Growing up in upstate New York, Carstensen was already a rebel. Even though her father was a college professor, she initially thumbed her nose at college and at age seventeen got married ("And I wasn't even pregnant," she says, still a bit amazed at her younger self). At one point, Carstensen got into a car accident, broke her leg, and ended up stuck for months in a rehabilitation center "with all the old women with broken hips."

And it was there that the seeds of insurrection were sown. Seeing that she was young and bored, the staff put Carstensen, then twenty-one, in charge of watching out for the older women, and as the months went by, she saw that some did well and some did not.

"So many of them had run out of money and were alone and had to sell their houses to pay for their care," Carstensen told me. "But others had a lot of family that came to visit and were the matriarchs of their families and were doing fine, and I began to question whether aging was just a biological process. It is biological, but it has to do with circumstances, with social context, even with emotions."

Carstensen became more intrigued by what she was seeing around her at the rehab center, and when her father brought her tapes of a psychology lecture class at a nearby university, Carstensen was hooked. "I didn't want to study medicine and just find out about the biology of aging. I wanted to know how biology and social influence interacted."

Once at Stanford, Carstensen set out to do just that, conducting study after study that looked at the intersection of aging and emotion to figure out exactly what was going on.

First, she tackled memory. In one of her most influential studies, published in 2003, Carstensen, along with Mather and psychologist Susan Turk Charles, at the University of California at Irvine, found that, starting in early middle age, around age forty-one, people recalled more positive images (smiling babies) than negative ones (ducks caught in an oil spill). They found that the shift continued for a number of years—they tested people up to age eighty—and applied equally to men and women, office workers and plumbers, and showed up consistently across ethnic groups.

"Older people clearly showed preference in memory and attention for positive over negative," Carstensen says.

Hints of this had been seen earlier. Some smaller studies, for instance, had shown that as we age, we remember and report more positive aspects of daily life. Asked about an apartment they'd seen, older people are more likely to first say something such as, "Oh, it had a really good kitchen," rather than, "The closets were way too small." As we get older, we report fewer bad moments from our days. And we're much less likely to label a whole day as bad just because of one untoward incident.

"It's not that people who are younger don't see the positive," says Susan Turk Charles, "but with younger people, the negative response is more at the ready. If you ask an older person what kind of day they had, they are more likely to say, 'Oh I had a good day,' and if you ask them if anything bad happened, they are much more likely to say no. But with younger people, it is the opposite; they are much more likely to say, 'Oh I had a very, very bad day. I had a big fight with my parents.'"

Aging Is the Answer

So, why this emphasis on the good side in life as we get older? Carstensen asked herself that very question. And after much consideration, a deep look at the literature, and more groundbreaking research, she settled on the answer: The shift occurs as we age because it comes from aging itself.

In the 2003 study "Aging and Emotional Memory: The Forgettable Nature of Negative Images for Older Adults," Carstensen wrote:

"Our research team has informally asked scores of older people how they regulate their emotions, particularly during difficult periods in their lives. Regularly, they responded with the answer: 'I just don't think about it [problem or worries].' At first this statement seemed to offer little insight into how older adults were regulating their feelings; however, the consistency of their responses made us turn to the possibility that processing positive and negative information may vary as a function of age."

The conclusion was not reached lightly. Rather it was based on a whole raft of studies that looked at the question from every conceivable angle. In science you don't always find a line of studies that progresses step by step, asking the next most logical question. But in the series of elegant studies, Carstensen and her colleagues did just that.

In one, for example, Carstensen wondered whether people didn't remember negative material as well simply because they ignored it altogether. But that was not it. Instead, the researchers found that if older people were presented with one image at a time, they looked at negative pictures even longer than positive ones, the same as younger people.

Then Carstensen and her team discovered another intriguing clue, zeroing in on choice. She found that even though we don't ignore negative information in middle age, if we are given the choice—positive or negative—we choose to focus more on the good than on the bad. Middle-aged people, for instance, were much faster at picking out small details on a happy-looking face than on an unhappy one.

Could it be, then, that negative images are simply much harder to process as we age, so, if we have the choice, we head toward the happier pictures because of a lack of energy, perhaps? Not at all. As Charles says, negative information is "much, much more potent" and remains much easier for brains to recognize and process. We actually have to work harder to focus on the positive.

"The literature is very, very clear on how much more potent the negative is," Charles says. "Even with rats, it only takes one bad thing, one shock or a bad taste in their food, and they will avoid that place or that food. It only takes one bad experience for them to learn. And it's the same with humans. If I see four friends and they all say, 'Boy, that is a great dress.' And then I see one other friend and she says, 'Boy, you really have put on weight,' guess which comment I will remember? And even in marriages, studies have shown that it takes *five* positives combined with one negative before someone will consider their marriage a happy one. If you have only two positives and one negative, that negative will wipe out the positive and people will consider the marriage a bad one. Believe me, the negative is much more powerful than the positive."

That means that even in middle age, our brains still register the bad things around us.

Okay, the researchers said, but maybe the part of the brain that responds to negative and threatening information—the amygdala—simply begins to wear out, so that no matter how potent the negative message, it doesn't register as strongly. But in further studies, Mather found that as we age our brains respond just as robustly to threats, a clear sign that our amygdalae are holding their own, even as we get older.

So what was behind this? What could be the reason for what Carstensen and her team began to call "the positivity effect," the increase in focus on the positive as we age? In the end, the researchers were left with only one real answer: We focus more on the positive as we age because we want to. It suits our goals and—though we do it without knowing we're doing it—we make it our business to sort out life this way.

And it is not that our brain gets lazy and wants to live out its days in some happy haze. On the contrary, Mather found conclusively that it's the best brains, the brightest brains, that have the most bias toward the positive.

And it might very well have to do with the least positive idea around—death. Carstensen believes that as we age we become much more aware that we have less time left in life—and it therefore becomes much more important for us to maintain emotional stability. One way to keep on an even keel is to steer clear of the bad and focus on the good. And, though we're not aware of it, we manipulate both our attention and our memory to suit that goal.

When we are young, negative information is paramount. We need to learn what to watch out for—the negative. But as we get older—and certainly by middle age—we already have a lot of cautionary knowledge. At that point, we may choose to gloss over a glitch here and there to focus on what's more important—regulating our emotions. And we do it because it's what we need—and want—to do.

"Time perspective is the dominating force that structures human motivations and goals," Carstensen says. "Humans have a conscious and subconscious awareness of their time left in life, and that perceived boundary on time directs attention to the emotionally meaningful aspects of life. When time is perceived as expansive, as it is in healthy young adults, goal striving and related motivation center around acquiring information. Novelty is valued and investments are made in expanding horizons. In contrast, when time is perceived as limited, emotional experience assumes primacy.

"When we are younger we orient toward the negative. When we are younger that information just has more value," she adds. "But increasingly with age, we see a shift. And I think it is because this shift serves to regulate our emotions. It's not that we are sitting around saying 'I will not focus on the negative.' It is not conscious. But it is not completely subconscious, either. I would say it is a motivated choice that we make because it is useful."

None of this means that at middle age we're in some blissful fog. If you, or someone you care about, has a serious setback, illness, or suffer from clinical depression, it's unlikely that you're at your most jolly, no matter what your age.

But the scientific findings have been remarkably consistent: Our middle-aged brains work incredibly hard to be enthusiastic about life, to see the good things—a trait that may be one of the biggest advantages a brain can have.

And the positive spin may have evolved because it works well for the species in general. There is a well-known thesis, sometimes called the "Grandmother Hypothesis," that postulates that humans and primates that had helpful, living grandmothers in their group lived longer. As Carstensen sees it, grandmothers with a brighter outlook gave the group a greater ability to thrive and survive.

"There is a powerful role in being calm and positive as we age; if older people are like that, it can help to keep the group together," Carstensen told me. "If you have strong negative reactions you might react too quickly and get too angry and that might not help. But if you have a grandmother who cares and is attached, perhaps the whole group will live longer. If that grandmother has an amygdala that allows her to be calmer . . . that might give everyone an advantage. It is cognition serving survival."

Emotional Regulation from the Frontal Cortex

There are hints, too, that the shift may involve more parts of the brain than just the amygdala—in particular, the frontal cortex, the region behind our foreheads that has grown huge in humans and helps us focus on what we want to focus on. In yet another clever experiment, Mather found that when she distracted older people, they no longer stressed the positive. That means that the part of the brain they used to deal with distraction, the frontal cortex, was distracted itself and could not help push attention toward the positive, even if that was what these older people, on some other level, wanted to do.

Other brain-scanning studies, too, show this in more detail. Joseph Mikels, at Cornell University, has found that older adults who emphasize the positive side of life the most also use their frontal cortex the most, in particular the section called the orbital frontal cortex, which has been linked to emotional regulation. In some cases, the amygdala may be able to do this on its own; in others, a healthy frontal cortex joins in to make sure it happens, which to Mikels is convincing evidence that "the positivity effect is regulatory in nature."

As Mikels himself confesses, this thought "goes against the grain—some of my students don't believe this, they say, 'my grandmother is the grouchiest person I know,' but then I ask them and they say, well, it's true she is lonely—and that's the reason."

But if our health and living situation are good, we gradually gain a brighter perspective because the structure and leanings of our brains start to head us in that direction.

"This is not a result of older adults wearing rose-colored glasses, but a function of their brains, which they have activated, and regulated, to focus on the positive and away from the negative," Mikels added. "We do it on some level on purpose. The ability to regulate emotions increases with age. This is one of the really good things about the middle-aged brain."

4 Experience. Judgment. Wisdom.

Do We Really Know What We're Talking About?

There's an argument to be made that the true test of a human brain is its ability to figure out other human brains.

Not long ago, when I mentioned that I was writing a book about the middle-aged brain to a friend, her first question was about the younger, trainee brains she had at home. As a mother of three girls, all in adolescence, she wanted to know, in a wishful way, only one thing: Does judgment improve? Do we get better at dealing with other humans, at making the right call?

Yes—and such insight is rooted in brain biology. We can now detect—even watch—mature judgment grow in our brains. The connections that help us identify the bad guys or the wrong road get stronger, and they may be at their strongest at middle age.

Thomas Hess, a psychologist at North Carolina State University, has done dozens of studies of what he calls "social expertise," which he finds peaks in midlife, when we are far better than those younger *and* older at judging the true character of others. Such tricky evaluations get easier—and closer to the mark—as we age. And it's the nature of how our brains develop that gives us that advantage.

By middle age, we not only have more years of experience with real people in the real world but the brain cells devoted to navigating the human landscape turn out to be exceptionally durable. Scanning studies show that parts of the frontal cortex that deal more with emotional regulation atrophy less quickly than other brain sections as we age. And it's that mix of emotional control, mental prowess, and life experience that helps us make the right calls.

"Some areas of the brain that appear to be involved in processing of socioemotional information . . . exhibit relatively less neuronal loss than other parts of the brain," Hess told me. "As individuals progress through life, they interact with others and acquire culturally based knowledge [about] . . . why people behave the ways that they do.

"The fact that middle-aged adults appear to be the most expert is consistent with notions that midlife is a time of optimal functioning," he added. "Basic cognitive abilities are still relatively high, and there's also a fair amount of experience . . . [so they] function at high levels in everyday settings."

And those everyday settings include a wide range of activities. David Laibson, at Harvard University, for example, has done fascinating studies in the emerging field of "neuroeconomics"—how people use their brains to make financial decisions—and he, too, finds we're most adept at this in middle age. Laibson has found that when confronting complex money issues, such as mortgages or interest rates, those in middle age make the best choices. In studies around the world, Laibson has found that people roughly between the ages of forty and sixty-five more easily grasp the consequences of financial decisions and have better judgment overall.

In fact, Laibson goes so far as to pinpoint the apex of all this: His research finds that those who use the best judgment in matters of personal economics are in their fifties.

"That seems to be the sweet spot in terms of all this," Laibson told me.

Weighing Wisdom

So what is this sweet spot? Is it judgment? Is it social expertise? Is this what we call wisdom?

The concept of wisdom—perhaps the most clichéd cliché of aging—has deep roots. It's mentioned frequently throughout literature, notably in the Bible, where it's described as a special mix of heart and mind. Most neuroscientists regard the concept with suspicion. Even now, those who will speak out loud about the idea divide into camps, albeit overlapping ones. Some assign wisdom's weight to emotional equilibrium, beginning with William James's famous declaration in 1890 that wisdom is "the art of knowing what to overlook."

Perhaps not surprisingly, there aren't many scientific studies that focus on the ability in middle age to keep one's eyes closed or mouth shut. But the James notion does have an uncanny similarity to work by Laura Carstensen and Mara Mather showing that emotional regulation increases with age.

As we get older, we also have more mixed emotions, a trait that works in our favor. A study by Susan Turk Charles found that when viewing a scene of clear injustice—a film clip from the movie *The Great Santini*, for instance, in which an African-American man with a lisp is mocked by a white man, or a clip from the movie *The Curse of the Working Class*, where a husband yells at and hits his wife— younger people react only with anger, but older people are both angry *and* sad.

This more complex, nuanced response to the world slows us down, restricting impulsive acts. And that may be good for our own survival, as well as that of the group—another case in which a middle-aged brain may function better simply because of how it's set up. "If you have one emotion it is easier to act," Charles explained. "And if you're on the savanna and a lion is chasing you, that quick action may help you get out of there. But in our complex world, it might be good to go slower, to think twice."

Even among scientists, the search for wisdom has a rich history and one not reserved to pure biology. One of the most prominent of the early life-span researchers, Paul Baltes, was, before he died several years ago, head of the highly respected Max Planck Institute for Human Development in Berlin. Baltes became fascinated with the possibility of scientifically deconstructing the building blocks of wisdom and spent years on what became known as the Berlin Wisdom Project. That project searched for wisdom anywhere it could, including the study of proverbs such as the Serenity Prayer ("God, grant me the serenity to accept the things I cannot change; courage to change the things I can; and wisdom to know the difference").

In the end, Baltes and his colleagues settled on a series of hypothetical questions about life choices, the right answers to which, they believed, equaled wisdom. The answers rested largely on the ability to consider variables—to look at the big, messy picture. For example, one question might be: What's the best way to get to Chicago?

Responding off the tops of their heads, some might answer quickly, saying something like, "Get on a plane."

But a few would take the time to consider the variables—the messy picture—and ask more questions to narrow the possibilities: "Well, tell me, how many people are going? How much time do you have to get there? How much do you want to spend? How long will you stay?"

And while such hypothetical questions might seem simplistic, they nevertheless illustrate the complex ways our brains operate day in and day out. Considering the various ramifications of a situation,

Baltes believed, means you have a brain that takes the measured, long—and wise—view.

After many years of such testing, Baltes and colleagues, while allowing that it's possible to be wise and young, decided that those who scored the highest on this sort of question and were, therefore, in their terms, the wisest were around sixty-five years old—and that peak was reached after a fairly long trek along the middle-aged "plateau" of sustained wisdom-ness.

Following in Baltes's footsteps more recently, Monika Ardelt, a sociology professor at the University of Florida in Gainesville, has put together an intriguing scale that determines how wise a person is by his ability to cope in the actual world. She measures a person's wisdom according to how well he performs in three dimensions: *cognitive,* which she describes as the "desire to know the truth and be able to look at gray and not see everything in black and white," as well as the ability to "make important decisions despite life's unpredictability"; *reflective,* the ability and willingness to look at different perspectives; and *affective,* the level of sympathy and compassion for others.

Ardelt has now matched outcomes on her measures against a set of data from Harvard University, which has been tracking a group of 150 men for more than forty years. Although she is still refining her findings, Ardelt told me that she's found distinct correlations between high scores on two different three-dimensional wisdom measures at midlife and in old age, and certain personality traits found in the Harvard study.

In an in-depth study of eight long-term participants, the most decisive factor that predicted wisdom was their level of self-centeredness. By her measure and Harvard's, it was those who focused on something outside themselves who turned out to be the most wise, a message, of course, that we've been told—and often ignored—for centuries.

"It was really striking," Ardelt told me. "Those who were high-high (wise at both fifty and eighty) also scored very low on self-centeredness. They cared about others. They were giving in some way or another. And those who were primarily concerned about themselves, or their standing in the community, scored very low on the wisdom scale."

Ardelt believes such wisdom comes directly from taking a broader perspective over time. Clearly, as she says, there are still "a lot of old fools" out there. Wisdom does not always develop automatically. And, as she puts it, we live in a society that, rather than rewarding those who are selfless—who teach or care for others—instead glorifies those who think mostly of their own gain—those who seek money for money's sake, for instance.

"We could have a society that fosters wisdom more," she said, a bit ruefully.

For the most part, die-hard neuroscientists have regarded this kind of discussion as squishy nonsense. But that's changing rapidly. Some are finding what they now call wisdom deep in the brain's very structure and workings—and in the midst of middle age.

In particular, many equate wisdom with an increased capacity, as we age, to recognize patterns and anticipate situations, to predict a likely future, and to act appropriately. As Neil Charness, who studies expertise, puts it, human brains are "pattern recognizers par excellence.

"Humans are not called homo sapiens sapiens—knowing man—for nothing," Charness says. "We can size up what is going on and figure out what course of action is most promising and we use hundreds of millions of patterns to guide the process."

John Gabrieli, the neuroscientist at MIT, says it helps to understand this signature talent by thinking of something as simple as an apple. Even if the apple is only an outline on paper, or painted purple, or has big bites taken out of it, we still recognize it as an apple because that's how our brains are set up. It might not look like a standard apple, but our brains, through years of building up connections, become quite good at recognizing even vaguely similar patterns and drawing appropriate conclusions. Studies have found that we handle situations better when we know something about the situation beforehand, when we recognize at least part of a pattern we've seen before, which is more likely to occur for a middle-aged brain than for a younger one. "It's stunning how well a brain can recognize patterns," Gabrieli says. "And particularly at middle age, we have small declines, but we have *huge* gains" in this ability to see connections.

In our own worlds, while we may take this for granted, we often have a sense that we can see patterns and grasp underlying concepts with greater ease. Elkhonon Goldberg, a professor of neurology at New York University School of Medicine, calls these established brain patterns "cognitive templates" and believes they're behind an older brain's ability to better predict and navigate life. Not long ago, Goldberg—at the "ripe middle age" of fifty-seven—decided to take stock of his own brain and the results were fairly good. Indeed, as he writes in *The Wisdom Paradox,* he began to realize that while he might have a harder time at strenuous mental workouts, he was also increasingly capable of a kind of "mental magic."

"Something rather intriguing is happening in my mind that did not happen in the past," he writes. "Frequently, when I am faced with what would appear from the outside to be a challenging problem, the grinding mental computation is somehow circumvented, rendered, as if by magic, unnecessary. The solution comes effortlessly, seamlessly, seemingly by itself. . . . I seem to have gained in my capacity for instantaneous, almost unfairly easy insight. . . . Is it perchance that coveted attribute . . . wisdom?"

If an older brain is confronted with new information, it might take longer for it to assimilate it and use it well. But faced with information that in some way—even a very small way—relates to what's already known, the middle-aged brain works quicker and smarter, discerning patterns and jumping to the logical endpoint.

A friend of mine who has been a doctor for more than thirty years said she can now often instantly evaluate a situation, making it easier to come up with effective solutions. "When I walk into a hospital room now, there's a lot in my head already," she said. "I can still be surprised. But in a lot of cases I can foresee what will happen and that helps a lot to figure out what to do, what will work best."

The Gist of It All

In many ways, of course, all this sounds a lot like what we like to call intuition or gut instinct. Neuroscientists don't like to use such words. They prefer the word *gist*.

Defined broadly, gist is the ability to understand—and remember—underlying major themes. Here again, we get better at grasping the big picture—because of the intrinsic nature of how our brains operate.

A series of intriguing studies has shown that we more easily wrap our brains around a main idea and remember it better, too, as we age. If you give a child a list of fruits—apple, pear, banana, grape, for instance—he will be quite good at reciting the list verbatim. But beginning sometime in our teen years—probably due to the natural pruning of little-used brain connections and a corresponding fine-tuning of our brains—we focus less on individual units and instead look at groupings. By middle age, we easily recognize broad categories.

"Verbatim memory begins to decline after young adulthood but 'gist memory' remains intact and gets better even into older old age," says Valerie Reyna, a neuroscientist at Cornell University who has done some of the most extensive studies in this area.

Another recent study along these lines found that as doctors gained more experience and became more accurate in making medical decisions about heart disease, for example, they made decisions, much like my friend the doctor, based less on a labored process of assembling remembered facts and more on gist—gut instinct—a shift that made reaching a conclusion both simpler and speedier.

"If you know a great deal about a topic, you can infer rather than remember," Reyna told me. "But, in addition, the nature of your reasoning, judgment, and decisions changes. You use gist to get to the bottom line more effectively, reducing the need to rely on memory for details."

In a way, it makes evolutionary sense for the brain to be set up this way. Confronted with vast savannas of stimuli, those who quickly brought all the stimuli together—odor, noise, movement— to understand the big picture would certainly have a better chance of surviving than those concentrating on tiny changes in the color of the leaves underfoot. Even in today's world, this talent proves handy. It serves us well—and studies back this up—to know from the get-go that a salesperson, for instance, is unlikely to give us the information we really need. We know we need to get a broader view. And as we age, we get better at looking beyond the obvious, in part because of how our brains develop.

"It makes sense as we age," says Reyna, "to rely on the part of our memories that is best preserved, and part of that is gist."

Linda Fried, dean of Columbia University's Mailman School of Public Health and a longtime expert in aging, says the abilities to see the vast canvas can foster creativity as well. We become more inclined to tie disparate threads together to make a new whole. "As you get older you can draw on objective knowledge and life experience and perhaps even intuition and they all get integrated and we can be more creative and solve complex problems that we could not solve when we were younger," she says. "I think we even get better at recognizing those complex problems to begin with. It's only when we are older that we have the patience and the strength and the willingness to go after the big core issues."

In fact, some have watched this sort of brain integration, or wisdom, with their own eyes—or at least the eyes of a sophisticated scanner. One of the most passionate of the current crop of wisdom hunters, George Bartzokis, a UCLA neuroscientist, believes that whatever we call this—judgment, expertise, wisdom, magic—it happens quite naturally as our brains move into middle age. And it may be what gives humans our edge.

A lively, self-confident Greek who spent much of his childhood in Romanian refugee camps before coming to America, Bartzokis remembers seeing nature documentaries as a child and wondering, Why are we so different from, say, chimpanzees? Since we share nearly 98 percent of our DNA with the chimp, our closest relatives, what makes the difference?

Our brains are bigger in certain areas, most notably the frontal lobes. But what is it inside a human brain that makes that brain region work so much better than a chimpanzee's? In fact, other animals—dolphins and elephants—have proportionately larger brains than ours. So what is going on?

Clearly, a large part of our human advantage comes not only from one brain part or another but also from the extensive system of connections—neural networks that build and strengthen, and allow us to keep a picture of, say, an entire air-traffic control system in our heads.

Insulating the Network

But while those basic networks—the gray matter—are crucial, it may be what holds those networks together—the white matter—that gives us our true advantage. No other animal has anywhere near as much white matter as we do. There are those, including Bartzokis, who believe it is the amount of white matter alone that has allowed us to develop such complex skills as language, for instance.

The white matter is made up of myelin—the fatty outer coating of the trillions of nerve fibers. The white matter acts as insulation on a wire and makes the connections work. Signals move faster and are less likely to leak out of a brain fiber that has been coated with myelin. This layer of fat, Bartzokis believes, is what makes the whole orchestra play together—and reach its cognitive crescendo—at middle age.

In a 2001 study, after scanning the brains of seventy men aged nineteen to seventy-six, Bartzokis found that in two crucial areas of the brain, the frontal lobes and the temporal lobes—the region devoted to language—myelin continued to increase well into middle age, peaking, on average, at around age fifty, and in some, continuing to build into the sixties. The study bolstered findings from years ago by scientists such as Frances Benes at Harvard who carefully measured the myelin of the brains she'd obtained from a nearby morgue. She, too, found that myelin continued to increase with age, and she, too, suggested that this might very well be what she called "middle-aged wisdom."

How could a coating of fat do that? There's little doubt that myelin is crucial in the brain. As a brain develops in childhood and neurons in the motor cortex are coated with myelin, the child becomes more coordinated, his hands more dexterous. When it starts to break down in diseases such as multiple sclerosis, for instance, a person can lose control of vital functions, such as balance.

The insulation allows the neuron to recover faster after signals have been sent and get ready to send the next signal more quickly, giving brain cells what Bartzokis calls "greater bandwidth," and boosting their processing capacity by an astonishing 3,000 percent. This essentially puts us "online" and allows a more integrated and comprehensive view of the world.

And this myelination does not happen overnight. It's a process. We build layers of myelin, and its architecture depends in part on how we use our brains. Myelin is produced by the glia cells in the brain, cells that cling to neurons and were for many years largely ignored by science. (Although there was a flurry of excitement a few years ago when, after Einstein's brain was examined, scientists discovered that he had many more glia cells than are normally found in the logic areas of the brain.)

At a certain point, a type of glia called an oligodendrocyte sends out long tentacles that begin to wrap the neuron arm, or axon, in the fatty myelin. The wrapping continues, creating what looks like links of sausages. We all progress at somewhat different speeds in this process of myelination. There's some evidence that females are better myelinators than males.

And recent studies confirm that myelin, while partly determined by our genetic blueprint, also thickens and becomes more efficient with deliberate use. As Michael Jordan was shooting basket after basket as he was growing up, for instance, it's very likely that his basket-shooting neurons got more and more coatings of myelin. More myelin means better brain signals—and better basket shooting, in his case.

"You can have all the dendrites [brain branches] you want, but you need to connect them—and for that you need speed and bandwidth, you need myelin," said Bartzokis. "This is what makes us human."

In some cases, small segments of myelin can start deteriorating in our forties—indeed, as a relatively late evolutionary add-on, it's particularly vulnerable to toxins. Its deterioration may lead to declines in cognitive areas. But through our forties and fifties and, if we are lucky and generally healthy, beyond, we also have an efficient myelin repair process. Until such maintenance breaks down, there's a net *gain* of myelin that continues well into our sixties, particularly in that crucial area, the frontal lobes.

This overall myelin buildup, Bartzokis believes, is the "brain biology behind becoming a wise middle-aged adult." A wise middle-aged *human* adult.

"It developed because it gave us an evolutionary advantage to have wise adults around who would not abandon their children to the lions," Bartzokis said. "The middle-aged in the tribe had learned to control their impulses and not send all the children off to be killed in stupid wars, for instance, and that made them better leaders.

"In a way," he adds,"we've always known this, but we're just showing it now in science. Look at the Constitution. It clearly says don't let anyone be president who is not at least thirty-five years old. The writers were not stupid. They looked around and said, 'Hey, we can't let anyone that young be president.'

"I'm fifty years old now myself," Bartzokis adds, "and I do find I look at things with a much broader view. I see the whole big picture easier. That's the formidable—the amazing—maturity of the middle-aged brain. That is wisdom."

5   The Middle in Motion

The Midlife Crisis Conspiracy

Our current version of middle age is new. In fact, the study of middle age is so new, as one scientist told me, "It's like researching nuclear physics, something that simply did not exist before."

Oddly, in recent years as we got this thicker slice of midlife, it was saddled with a sour taste. Although the initial explanations did not necessarily use the language or tools of brain biology, they nevertheless attempted to characterize the state of the middle-aged mind. And that state, according to early conclusions, was not a happy one. For reasons that still baffle me a bit, news that should have been greeted with hope—longer life spans with more time in the middle—instead seems to have sent us into a tailspin. It was not just John Updike and Gail Sheehy. They got their signals from such scientists as psychologist Erik Erikson, who decided that to move from one stage of life to another, we had to undergo a bad and unsettling psychological crisis.

The Midlife Crisis

Then, to give that idea wings, came Elliott Jaques, considered the Father of the Midlife Crisis. And it's not that he was *having* one. In fact, Jaques had a long and distinguished career as an industrial psychologist, known for his detailed studies of human efficiency. But, almost as a sideline, he noticed that artists—at least an arbitrary sample of artists he studied—seemed to change their styles as they reached the midpoint in their lives, with some painters shifting to a more somber tone. To him, midlife, with its growing awareness of mortality, brought mostly a deep sense of loss and depression.

"What is simple from the point of view of chronology, however, is not simple psychologically," Jaques wrote after concluding his artist study. "The individual has stopped growing up and has begun to grow old," adding his belief that it is the "inevitability of one's own eventual personal death that is the central and crucial feature of the midlife phase."

Hard as it is to believe, Jaques's small study of a few randomly selected artists in 1965 seems to have spawned a near cult following of the idea of the midlife crisis, a notion that entered the popular culture thanks not only to Gail Sheehy in *Passages* but also to former Yale psychology professor Daniel Levinson, who, in his book *The Seasons of a Man's Life*, talks of his own self-styled study of middle-aged men.

"A man at mid-life is suffering some loss of his youthful vitality and, often, some insult to his youthful narcissistic pride," Levinson wrote. "Although he is not literally close to death or undergoing severe bodily decline, he typically experiences these changes as a fundamental threat. . . . Dealing with his mortality means that a man must engage in mourning for the dying self of youth . . . he must experience some degree of crisis and despair. . . . For large numbers of men, life in the middle years is a process of gradual or rapid stagnation, of alienation from the world and from the self."

Levinson's book, published in 1978, however, was based on only forty men, specifically selected by Levinson himself. From that tiny sample, Levinson located what he called this "Mid-life Transition" somewhere from age forty to age forty-seven, concluding that "for the . . . majority of men . . . this period evokes tumultuous struggles within the self and with the external world. Their Mid-Life Transition is a time of moderate or severe crisis. Every aspect of their lives comes into question, and they are horrified by much that is revealed. They are full of recriminations against themselves and others."

But, while Levinson is still read today and movies and magazine articles about midlife crises are still being written, in academic circles the idea has long been discounted.

Indeed, as results from long-term studies have begun to roll in, the picture of middle age has been flipped upside down, and the idea of a predictable or common midlife crisis—however much it is a part of popular thought—has turned out to be a myth. More rigorous research has painted a portrait of aging in general, and middle age in particular, that is very different from our widely held beliefs.

Even more important, with new tools, we can now look inside our own brains to see what's actually going on as we think, feel, and age. We can watch our amygdalae, our cortices, our hippocampi, in real time.

And as Stanford psychologist Laura Carstensen puts it: "There is no, absolutely no, empirical evidence for a midlife crisis."

In 1999, for instance, one of the biggest, and at the time only, studies of middle age, the MacArthur Foundation Research Network on Successful Midlife Development, which is still ongoing, found no evidence that crises occurred more frequently in midlife than at any other age. In fact, the ten-year study of nearly eight thousand Americans found that only 5 percent reported any kind of midlife trauma, and they were, by and large, people who'd had traumas throughout their lives.

Instead, between the ages of thirty-five and sixty-five, and, in particular, between the ages of forty and sixty, people across the board reported increased feelings of *well-being*. Women said they found menopause not the sea of sweat and sadness that it had been portrayed as but a "relief." Most felt they were productive, engaged in meaningful activities, and had a greater sense of control over their lives, including their marriages, which were also relatively happy.

It's true that most at midlife acknowledge a fair amount of stress. But in a finding that goes against what we thought we knew about middle age, most people in these more sophisticated larger studies say they are not only coping with the stress but that the coping itself makes them feel good about themselves. As one researcher put it, by midlife, we are "equipped for overload."

"The reason why midlife people have these stressors is that they actually have more control over their lives than earlier and later in life," psychologist David M. Almeida, now at Penn State University, said when the MacArthur results were first reported. "When people describe these stressors, they often talk in terms of meeting the challenge." Summing up, Harvard's Ronald Kessler, a director of the middle-aged survey, said simply, "The data show that middle age is the very best time of life."

And the good news doesn't stop there. More recently, another smaller study, which tracked the lives of a group of women who were seniors at Mills College in California in 1958, came to similar conclusions. In 2005, researchers Ravenna Helson and Christopher J. Soto at the University of California at Berkeley reported that after gathering more than forty years of data, it was clear that as the women moved into middle age, their moods got better, not worse. At the same time, according to Soto, they also "became more confident, assertive, and responsible."

The women had higher self-esteem and their moods—as well as their ability to regulate emotions overall—seemed to peak at around age fifty-two and hold steady for quite a while after that. What's more, as their children left home and the women had more time on their hands, far from rattling around dejected from empty room to empty room, they "took advantage of this time to do new and interesting things," Soto says.

Still a relatively young researcher at age twenty-nine, Soto told me he found all this quite heartening. "In my generation we have grown up in this culture that highly values youth and there are these markers that show you that you are over the hill," he said. "So it is good to see that when you actually look at real lives they continue to get better and better into middle age."

Soto also admits that if he weren't so dedicated to finding out such things in such a highly scientific manner, he could have gotten a hint about all this from his own mother, who, at age fifty-seven, is having a grand old time. After raising three boys, she went back to school in Spain, got her master's degree in Spanish, started teaching at the high school and the local college, and, along with Soto's father, is more active socially than she ever was before. Both have been lucky enough to retain their general health and their minds. "Maybe because people are not only living longer but are in so much better shape physically and mentally, lives just get better," Soto said.

And it's not just women. In a twenty-two-year study of nearly two thousand men that ended in 2005, Daniel K. Mroczek, a psychologist at Purdue University, found, after controlling for health, marital status, and income, that life satisfaction actually peaked at age sixty-five.

For his part, Mroczek also buys the theory that as we age we get happier largely because our brains learn how to regulate our emotions better. "Frankly, I think we just rewire our brains as we get older," he says. "You learn to handle things. It's related to time but it's unconscious. Your brain decides, on some level, to look at the world differently."

There's no question that what we do—how we live—alters our

brains. Although for years it was taken as gospel that the brain was largely fixed by adulthood, that gospel has been dispelled. Ever since a Canadian researcher took his adult lab rats home to run around in his house and later found that those rats were considerably smarter than the rats left behind in their boring cages, neuroscience has systematically upended the idea that the adult brain cannot change its structure or improve how it works. It can and does. What we do changes the architecture of our brains. It's called *neuroplasticity* and it's the underpinning of everything we now know about the brain.

Both animal and human brains are plastic, mutable. Experiments with rats, dogs, and monkeys have found that those in "enriched and stimulating environments" (for a rat that means toys and rat pals in their cages) wind up with bigger brains, more connections, and are, on every test imaginable, much smarter than those living lonely, mundane lives.

In humans, too, there's now ample evidence that the adult brain reorganizes and continues to develop. Our brains have evolved to be as nimble as possible. Since brains do not necessarily know what type of environment they'll find themselves in as they go along, they have to be able to adapt to survive. Now-famous studies of London taxi drivers and expert violin players found that areas in their brains devoted to spatial reasoning (taxi drivers) or fingering strings (violin players) grew larger as the drivers drove through London streets or the musicians played. We'll come back to this idea, and the larger topic of neuroplasticity, in part 2.

Whose Empty Nest?

In some ways, it's easy to see how we got the picture of middle age wrong for so long. As I've mentioned, we had never encountered this particular beast before. In fact, that other great myth of middle age—the empty nest syndrome—is also now considered to be largely fiction.

Every year, Karen L. Fingerman, forty-one, a psychologist also at Purdue University, starts her lectures by asking incoming freshmen how they think their parents are doing now that their children are in college. Every year, the students say they must be "devastated" by their absence.

But nothing could be further from the truth. When science actually looked at how parents were doing when their kids left, researchers found that they were doing just fine.

Again, I admit that I fell for this one, too. Not long ago, I went to a meeting of a book club I've been a member of for years. Most of the women in the group had kids who were a few years older than mine and had recently left for college. I expected to find the group in a deep funk. But instead, the women were complaining about how *often* the kids were calling them. "I had to tell my daughter to stop calling me at work, that I have work to do," one woman said. They missed their kids and loved seeing them when they were home, but because the kids were, as one mother said, "doing what they're supposed to be doing," the parents felt good, not bad. They were proud—and busy themselves. "Empty nest syndrome? Ha," said one. "I miss them, sure. But I have a *lot* to do."

In fact, in her own research, this is exactly what Fingerman, along with others looking into this area, has discovered. Studying mothers and daughters, Fingerman found that both seem to do better once they are under different roofs, even if one is a dormitory roof. They're happy in what they're doing and, with all that "teenage tension gone out of the relationship," much happier with each other, too. "Some girls in their early twenties actually wax poetic about the good relationship they have with their parents," Fingerman says.

Certainly, there can be a sense of wistfulness as we find our houses without the high energy and verve of growing teenagers, but the empty nest *syndrome* was simply made up because popular culture somehow needed, as Fingerman says, a "female counterpart" to the midlife crisis, which was geared mostly toward men.

But no one has ever been able to find a true empty nest syndrome in a scientific way. Instead, even among women who devote all their time to raising their kids, studies find mostly a "great deal of satisfaction" when the kids become independent. "They feel they have done a good job and they suddenly have the freedom to do new things. They feel great," Fingerman says.

There is a tendency in psychology to pigeonhole all manner of life events. But why some events are singled out and not others is a mystery. When I spoke with Fingerman, she had just dropped off her child at kindergarten for the first time and wondered why there wasn't a syndrome name for that: The "First Kid in Kindergarten Syndrome." Or, having overheard her colleagues fret about their newly licensed sixteen-year-olds driving in the winter, she wondered, "Why don't we have a label for that, the 'Parents of Kids Driving in the Snow Syndrome'?"

In fact, at midlife volatility diminishes because more is settled. When she looks at her own students in their twenties, Fingerman says, she sees considerably more depression and upheaval than she sees among her own age group.

"They are more emotional and things are just bigger in their lives," she says. "It's natural. Things can go badly for me, maybe I don't get a grant funded, but, look, I have tenure, I still have a job. For these kids there are so many unknowns and bad things can seem so big. As we get older, I really do believe we get better at emotional regulation. I can learn to avoid high-stress situations. We just get better at that."

There are other benefits of the empty nest that I had not even thought about myself, but, on reflection, are true, too. Victoria

Bedford, a psychologist at the University of Indianapolis, has found that as children leave home, parents often find a bright spot none ever expected: They reconnect with their siblings. It's as if, after living hectic lives, people look up and say, "Hey, I know I have a brother out there somewhere."

As Bedford says: "This is true with men and women. Sibling relationships are always important but when the kids are gone you are more settled and have time to connect with your sibling. And we've found that people actually do."

As I thought of this, I realized that that, too, had happened in my own life. I had lived for many years in New York, far away from where I grew up in California, and my older brother was in Phoenix. We'd see each other once a year, but he was busy as a teacher and a coach and I was busy with newspapering. Then, as often happens, when our father was dying in California, my brother and I spent days and nights at the hospital, then cleaning out and selling a house. I was reminded that my brother is one of the nicest and funniest people I know and realized how much I missed him. Since then, we e-mail all the time, he calls my daughters, and we try to visit each other more often. Since I last knew my brother best when we fought in the backseat of the car on family trips, this grown-up sibling relationship has been one of the nicest things that has happened in my middle-aged life.

Bedford, at sixty-one, reports that she has no empty nest doldrums and doesn't know anyone else her age who has them, either. One of her daughters, she says, was a "terrible teenager" and now they have a "wonderful relationship. Your children become more like peers and it is great," she says. "There is no question that I am happier now in midlife. And I have not seen anyone upset about an empty nest. Not in the least."

So why did earlier depressive ideas persist for so long? Were theories of midlife crises, with their red Porsches and pretty young things, nothing more than, as one researcher insisted, a "collective fantasy by white men *for* white men"?

Carol Ryff, who leads the MacArthur Foundation's middle-age survey, says that the midlife crisis probably applied to only a narrow group of men, during a narrow span of time—those who had returned from World War II, rushed into family, house, and kids, and then, having caught the 7:05 train for twenty years, woke up and wondered what they were doing.

"Maybe all this was true for a very small group," Ryff told me. "These are men whose lives were disrupted. They went off to fight the war that interrupted their careers and their pursuits, and then they came back and jumped quickly into things. They felt they had to catch up with marriages and jobs. And then, when they reached fifty, they looked up and said, 'Hey, is this the life I really wanted?'"

In fact, it's surprising how flimsy the evidence—scientific, biological, or empirical—is for these strong beliefs that have crept into our life narratives. The first reference I could find to the empty nest syndrome was from a small pilot study published in June 1966 in *The American Journal of Psychiatry*. The study, called "The Empty Nest: Psychological Aspects of Conflict Between Depressed Women and Their Grown Children," was based on sixteen women whose children had left home and who were depressed. And the study's authors decided there was at least a "temporal relationship" between the two.

"In our depressed patients, it appeared that cessation of child rearing influenced the content of symptoms presented and that in cases of women who had ended child-rearing functions, there was almost always some degree of conflict between them and their adult children.

"Based on these observations, we undertook to study the empty nest syndrome, defined as the temporal association of clinical depression with the cessation of child rearing," the authors wrote.

These patients, however, were not only a tiny group but were also hardly representative. For one thing, they were all sufficiently depressed to be hospitalized in the Massachusetts Mental Health Center. Those with the most severe symptoms, as the researchers conceded, were all "foreign born or first generation Americans who clung to the traditions of their countries of origins and had achieved a median of only nine years in school. They had married while still teenagers and had started their families almost immediately and were socially very withdrawn having few or no friends or interests outside the home."

From that skewed sample, which the authors themselves said provided proof of nothing, was hatched this empty nest idea that has refused to die—for decades. The first mention of it that I could find in popular literature appeared in a 1972 article in *Ladies' Home Journal* in an interview with Pat Nixon in the White House, who was shown modeling spring fashions. The profile said that Pat "likes to take a positive approach," and that "Even the empty nest syndrome hasn't seemed to hit her. She seems genuinely delighted by both of her daughters' marriages." Pat Nixon, in her own way, was simply voicing what science is now confirming. In middle age, if we're lucky to be healthy and alert, we start to take a decidedly positive approach—just like Pat.

Youth Culture's Taint

But old myths die hard and our current culture is not helping, either. Richard A. Shweder of the University of Chicago, in trying to explain why we got such a dour view of our path through life, puts considerable blame on our Western society, which continues to stress "physical and mental decline."

In a wonderful book of essays called *Welcome to Middle Age! (And Other Cultural Fictions)* he points out that in other cultures middle adulthood is not defined by "back pains," but is instead marked by increased status as people gain "family position and social responsibilities." In many Hindu households, Shweder says, referring in part to his own work and to that of Usha Menon of Drexel University, there is not even a concept of midlife and no word for it. Instead, they have the word *prauda,* which means "mature adulthood, which begins whenever a married woman takes over the management of the extended household and ends whenever she relinquishes control and social responsibility to others."

One of the most interesting of Shweder's essayists, Margaret Gullette, a resident scholar at Brandeis University, believes that in our culture we remain victims of the "ideology of decline" that is "raining over us."

We have allowed ourselves to be "aged by culture," taught to think of our lives in simply an "age graded" way, based on the misguided sense that "the body fails at midlife and this bodily failure matters more than anything else," while the positive aspects of aging, "maturity, competence, compassion, etc.," are not "coded as age associated."

As she sees it, such views persist primarily because they serve well those who want to keep us buying "wrinkle creams."

"Midlife decline ideology," she writes, works to "enclose us in self doubt, embarrassment, shame, humiliation, despair. It fosters narcissism. By learning to concentrate on an 'aging' body, the twentieth century midlife subject learns how isolated and helpless he or she is."

And while this might sound a bit harsh, it makes perfect sense to many of the people I spoke with who are going through middle age in the twenty-first century.

Susan Nowlin of Bloomington, Indiana, for instance, says that as she reached middle age it seemed as if the world had a preset agenda for her: When she turned sixty, she thought she should retire as a teacher because that was what the "culture was pushing." But then rules were changed in her district and she could not retire—and it was the best thing that could have happened to her.

At sixty-one, she is still teaching English to 150 middle school students in Bloomington. She walks thirty-five minutes a day, recently bought a DVD so she could learn weight lifting, plays bridge, and is planning on writing a novel. And when she is not doing any of those activities, she goes out to dinner and a movie with her husband, sixty-three, an industrial psychologist, and they, she says, "laugh a lot and have a ball."

To Nowlin, the concepts of midlife crises and empty nests have proven to be overblown and may just "have applied to a few people." When her two sons left home, Nowlin says she did have a bad moment. She recalls standing in the dining room one evening and "feeling a bit lonely." But that was it for her empty nest syndrome. Now she finds she is not only busier than ever but also more optimistic and calmer. And she finds that taking a broad view often helps.

"I see some younger teachers and they harp on the smallest things and destroy their relationships with the kids," she says. "Now every day I think, Have I done something to help a kid? And if I have, I really feel like I accomplished something. Maybe that is why I am so happy."

~~~ Part Two: The Inner Workings

6   What Changes with Time

Glitches the Brain Learns to Deal With

I was talking with Deborah Burke, a neuroscientist at Pomona College in California, and we were having one of those maddening conversations in which neither of us could remember the name of the person we were talking about.

She was telling me about a dinner she'd been to the night before where she could not, for the life of her, remember the name of a well-known scientist.

"I was at this party and I was talking and I just could not remember Richard Dawkins's name," Burke was telling me. "Then I couldn't remember the name of that evangelical guy, you know, the one who was against homosexuals and then turned out to be one? Oh, you know. What was his name? Oh, dear, now it's happening again."

I tried to think myself. I wanted to help. I could see this guy in my mind, picture him giving a news conference. I, too, knew his name well. Just not at that moment.

"I think his name was Ted. Ted something. And I think it had an *H* in it," I offered, doing the best I could.

In fact, over the course of our conversation—for the next *two* hours—neither Burke nor I could come up with this man's name. At certain points, Burke would stop and wonder out loud, "Oh, what is his name? This is really bugging me. I can see his church. It's like a big warehouse. What *is* his name? I'm going crazy."

We were not going crazy. We were mired in the Swamp of Lost Names.

And this time, I was mired along with one of the leading authorities on why we lose names to begin with. In particular, Burke studies what we call "tip of the tongue," that gnawing feeling that you *know* something but you just can't find it in your brain, a sensation likened to being on the brink of a sneeze.

Over the last few years, Burke has tried to figure out why, in fact, we can't sneeze. Where are those names anyhow? If they are on the tips of our tongues, why can't we spit them out?

Despite all the newly recognized powers of the brain in middle age, there are—as we all know—a few glitches as well. The truth is, by midlife, most of our brains show some fraying around the edges and names are often the first edge to go ragged. Even names we know well vanish. We have the strong feeling we know the name, but we just cannot bring it to mind. Who are you?

It's annoying. It's frightening. In her research, Burke has found that tip-of-the-tongue incidents, or "Tots," as she calls them, start to creep in as young as thirty-five and are a big part of middle age. Tots are much more common with proper names than with the names of objects. And they also occur more often with proper names than, say, the names of occupations.

"If I say Mr. Baker is a potter and Mr. Potter is a baker, you will remember the occupation, that someone is a baker or a potter, much easier than the names Mr. Potter or Mr. Baker," Burke told me. Indeed, when I see my plumber, my brain easily says to me, "There's my plumber," but his name . . . well. That's not popping into my head.

In survey after survey, this tip-of-the-tongue phenomenon is listed as the most irritating, embarrassing, and worrisome part of the aging brain.

So why does it happen? After all, remembering names—Mama, for instance—is one of the first vocal tricks we learn and seems fairly crucial to the species. Why lose that? As we age, our overall vocabularies improve. As Burke says, "A seventy-year-old has a better vocabulary than a twenty- or thirty-year-old. We just keep accumulating words and our verbal abilities get better."

So with all those words and stacks of names stuffed inside our heads, why can't we reach in and grab the one we want? Do we have too many names in there? Do they get lost in the clutter—or just lost?

A Retrieval, Not a Storage, Problem

Well, for starters, the names are not technically gone. Research into the cellular activity of the aging hippocampus—where most memories are processed—indicates that much of what we learn, in the form of chemical markers, is not missing, it's just at the bottom of the pile. For the most part, it's a problem of retrieval, not storage. It's like trying to find the right book in a well-stocked library.

Burke's work has shown this elegantly. In one study, she found that if an older person is shown a picture of someone well known—say, Brad Pitt—and has the feeling that he *knows* the name but can't recall it—a Tot—he will be much more likely to retrieve the name successfully later on if he is, in between, asked to answer a question whose answer is "cherry pit." Even though he is unaware that the mention of *cherry pit* is in any way connected to the picture of Brad Pitt, the twinning of the sounds in the two words—*pit* and *Pitt*—is enough to prime the memory on a subconscious level and help him get over his tip-of-the-tongue problem and retrieve that name. Interestingly, such prompting does not generally improve performance with young brains, but does help middle-aged and older brains, presumably because that's when we need the help.

Burke has also discovered why lost names often come to us, seemingly out of the blue, later on, usually long *after* we need them—again because of certain clues we're unaware of. If you're trying, unsuccessfully, to recall the word *Velcro,* for instance, and you later hear the word *pellet,* it's much more likely that the heretofore-missing word, *Velcro,* will spring to mind. Even when it's only an internal sound that is similar—in this case, "el" occurs in both words—the second word can draw out the lost word. To us, it seems the words come out of nowhere. Burke calls these "pop-ups." And they, too, occur more often as we age.

But if our brains are doing so well in middle age, why do names go missing in the first place? Burke's theory is that "it's because of the way words are stored and organized in the brain." She says, "The sound of the word—its phonology—and the information about that word—the concept of the word—are in different areas of the brain and the connection between them weakens. It can weaken if we don't use the name. But it also weakens as we age," much like that running muscle you haven't used much lately, either. It happens most often with the names of people we know but have not seen recently.

It happens, too, because the link between a person and his name is so arbitrary. Names that are unusually descriptive, like Grumpy, or names that have acquired meaning from the characteristics of a person they refer to, such as Scrooge, are remembered more easily than random names such as Peter Pan. (This must be why I've never forgotten the name of my childhood dentist, Dr. Smiley.) In general, there's absolutely no reason for Brad Pitt to be called Brad Pitt. There's no reason for Mr. Baker to be called Mr. Baker.

At the same time, we remember what a person does. That's because a person's occupation embodies a wide range of information that's stashed all over the brain, and that can be retrieved through various paths. When we hear that someone is a baker and later we're trying to come up with that fact, we might get there through an assortment of associations, from *white* and *apron* to *flour* or even *hats*.

"If I say *baker,* all sorts of information is called to mind," Burke says. "There are different ways to activate the 100,000 neurons, lots of different connections that lead you to that concept."

The thing to remember—if possible—is that forgetting names is part of normal aging and it is only one piece of processing an identification or recognition. If you forget that your husband's boss's name is Ed, it might be a bit embarrassing at the office party. But it's not Alzheimer's, a progressive disease where you might forget you have a boss, or even what a boss is.

At age sixty, Burke says she doesn't spend much time fretting about all this and she doesn't think the rest of us should, either. But it might help to plan ahead. Before going to a party, Burke sometimes makes a list of who will be there. She also uses a trick many of us secretly use. If she meets someone she knows but whose name she has forgotten, she resorts to the alphabet, going through each letter until she gets to one that prompts the name.

Still, it's unsettling. At the end of our talk, Burke still could not remember the name of the evangelical pastor. As we spoke, she'd interrupt herself to say, "This is driving me nuts."

"That is the emotional part of this," she said. "It can make you very upset. What was his name?"

At that point, I decided I'd better do something. While I was talking with Burke on the phone, I stopped taking notes, went to Google, and typed in "evangelical, resigned, homosexual."

That's the other lesson here, of course. As we learn to love and accept our middle-aged brains, we should—rather than panicking over these little peccadilloes—relax and get help. We're lucky. Those of us in middle age now are the first group to have a neurological elf standing by—the World Wide Web.

And, of course, in a second I had our answer. "It was Ted!" I told Burke, feeling triumphant. "I looked it up and his name was Ted and it did have an *H*. Ted Haggard."

Together, we sighed in relief. "Ah, that's it," Burke said. "I just knew it was Ted something. Thank you."

A World of Distractions

And so we both felt a bit better. But—as we all know, too—it doesn't end with a name here and there. By middle age, if our brains are not misplacing names, they're often misplacing themselves. We get distracted.

And it doesn't take much to knock us off course. A doorbell rings and we forget we're boiling water for the potatoes. We meet a friend at the hardware store and, after a brief chat, we no longer remember we went to the store to get a rake. My friend Phillis, fifty-one, who runs her own consulting company, told me that as she was climbing her building's stairs to go to her office on the fourth floor not long ago, she suddenly looked up and found herself on the eighth floor. As she climbed, she had glanced out a window to see who was in her parking spot, got distracted, and simply went right by her own office door. "Oh, my goodness," she said, "I've never done that before. Is that middle age?"

Such wanderings become increasingly common as our brains age. "When I ask my patients what is troubling them, distraction comes up again and again," says Adam Gazzaley, a neurologist at the University of California at San Francisco. "They'll tell me that they're sitting on the couch and they go to the kitchen to get something and by the time they get there they don't remember what they went there for. I hear that all the time. And when I ask them how that occurs, they say that something distracts them, maybe the phone rings, something makes them not pay attention."

Not long ago, the writer Judith Warner, talking about her new distractibility, confessed that she had "invited a couple to dinner and forgot to give them our unlisted phone number or address" and sent her daughter "to dinner at another family's house and neglected to tell anyone that she was coming."

Later on, Warner was heartened when, undergoing an MRI for migraines, her mindlessness was explained. She had a "hole" in her head. Her neurologist told her it was an inconsequential "small cystic area," but its very existence, however unrelated to any brain difficulties or maladies, was reassuring, at least to Warner, who, perhaps only half kidding, wrote:

> The self-blame game is now over. I no longer have to feel ashamed when—despite my ability to recall the details of a small news item from six years ago—I cannot remember the name, or even the face, of a person I met earlier in the day. No one has the right to laugh at me anymore when I write down important reminders—12:30 dismissal! Bring napkins!—on the palm of my hand. For I have a hole in my brain.

So, is that it? Do we all just have holes in our brains? Losing names is one thing, but losing entire dinner plans? If our brains are capable of so much at middle age—such expertise and wisdom and clarity and optimism—why do we walk right past our own office door?

Memory is a strange phenomenon and not completely understood on the molecular level. (One part likely involves the astonishing capacity of brain cells, sometimes described as "soft cells," to physically change their structure as needed. As it's often said, "brain cells that fire together, wire together." If two brain cells are activated at the same time, they will actually change their structure, form stronger connections—and let us form memories and learn. That means, for instance, that if you see a red bird and hear its song enough times, the neurons that recorded the sight of the bird and the neurons that registered the bird's sound are linked and physically altered. And the next time you hear that song, those neurons will fire up more or less in tandem and you'll think, "Hey, it's that noisy red bird again!")

While there remains considerable fussing over how this works, it is clear that memory is not a single mechanism. Names are arranged one way, plans for an upcoming dinner party another, and that vivid picture of the giant dog that chased you down the street when you were four years old yet another.

And by middle age, most memory functions—I'm happy to report—*are* still humming along nicely. Biographical material, for instance, generally stays intact. You remember who you are and who your brothers and sisters and cousins are, where you went to second grade. Even personal information that's acquired in middle age or later generally stays put. You know where you last worked. You remember how to make oatmeal, where the milk is. You can still ride a bike and drive a car, and you can, if you practice enough, perfect your tennis forehand—motor and muscle memory remain intact.

Episodic Memory

But other, more complex types of memory get a bit dodgy. Take a short break from a book you're reading, even for a day, and you'll forget not only what you've read on the last few pages but that you've *read that book at all*. A friend, Michael, told me that on a recent plane ride, he settled in to finish a book he'd started earlier that same week. But after he picked up the book, he found he "couldn't remember ever reading any of it." Unwilling to admit he'd forgotten what he *knew* he'd just read, he decided to start the book in the middle anyhow. "I just started reading halfway through," he told me. "I have my pride."

Such recollections for recent events—books we've just read, breakfasts we just ate—are called episodic memories. And our talent in this area generally does not blossom with age.

Why? How can some forms of memory parts stay put while others go missing? Do we, by middle age, simply have so many meals and movies and books in our heads that we have to get rid of some—a storage issue? It's true that our brains have to jettison something or we'd explode. In fact, the few people who throughout history have been incapable of forgetting anything have been driven crazy as a result. Our brains are set up to set priorities, to weed out the irrelevant.

Still, you'd think the basic outline of a book you're enjoying would stay put. Could we simply have too many weighty matters on our minds in general? Maybe we just can't be bothered using up valuable brain space to remember what was on pages 1 through 67?

Marilyn Albert, a neuroscientist at Johns Hopkins University who has been studying the aging brain for decades, says that some difficulties in the normal healthy brain are not imaginary—and not a simple issue of overload. "We used to think it was because we had too much on our minds or because we have been away from school for so long," Albert said recently. "But the declines are real and they begin in middle age."

In fact, our increasing problems with some complex types of memory can be tied to how our brain changes its functions as it ages. And researchers are now able to see how this happens.

Cheryl Grady, a brain scientist at the University of Toronto, for instance, has actually watched the middle-aged brain take a few detours into distraction. Using a brain scanner, she has caught it daydreaming.

In a recent study, Grady found that the key part of the brain that we use to concentrate—the dorsolateral prefrontal cortex, part of that crucial frontal lobe region—lights up red-hot, as expected, in young adults when they're asked to recall words or pictures they've just seen—a kind of difficult-to-do episodic memory.

But by middle age, she finds, such focused thoughts can be shoved aside by just about anything. As she scanned the brains of study participants, Grady was surprised to find that many older people trying to recall more complex information used their key frontal brain areas a bit less and a lower section of the brain more. And this second area is not helping. In fact, this fascinating brain region, called the default area—a region whose recent identification is one of the major discoveries in how the brain operates—is a key to why middle-aged brains can sometimes find themselves drawing a blank.

"This is the region we use when we're thinking about ourselves, our internal monologues," Grady told me as she explained her recent findings. "For instance, if you're in a brain scanner and you aren't doing anything, you might be thinking, 'Gee, I'm kinda uncomfortable.' Or you might be thinking that you should get some milk at the store later on that day. This is the part of the brain that we call the default mode. It's what the brain uses to daydream."

Starting in middle age, the brain's ability to switch *off* the default mode starts to wane. Faced with the task of remembering we're boiling water, our brains veer off into their own internal worlds, thinking about those great boots we'd like to buy or that football game we're planning to watch, none of it pertinent to the task at hand. And while we muse, all thoughts of boiling water disappear.

"This is one of the areas in which the aging brain does not do so well," Grady told me. "Our ability to tune out irrelevant material is reduced. In middle age, we seem to be in transition from the patterns in youth to those of older age in this area. And it might be one of the reasons we become more distractible."

Power to Focus

In fact, the ability to focus is one of our most crucial brain functions. It's a skill we acquire as babies and hone throughout adolescence. And it depends, to a large degree, on the development of our frontal lobes, which are not fully matured until we're twenty-five years old. This area helps us to focus, in part by blocking out—inhibiting—irrelevant details.

In a recent study using functional MRI, which can observe activity in the brain, Adam Gazzaley has also watched older people have more trouble keeping their brains focused. Shown both faces and scenes and told to focus only on faces, they had more activity in the area of their brain that registers faces—appropriately. But the area that registers scenes, which should have been suppressed or inhibited, also became active. And the older adults who had the most trouble focusing also had the most trouble remembering what they saw.

As we age, our frontal lobes don't block out irrelevant details that interfere as well, perhaps because they switch into default mode, or because of declines in connections or in the brain's chemical messengers, creating what's called an "inhibitory deficit." Explaining their own recent findings, published in the journal *Nature Neuroscience* in 2005, Gazzaley and coauthor Mark D'Esposito, a professor of neuroscience at the University of California at Berkeley, concluded: "older individuals are able to focus on pertinent information but are overwhelmed by interference from failing to ignore distracting information."

When I spoke with Gazzaley, he had just finished another scanning study that tried to pinpoint exactly when this happens as we attempt to pay attention. Not only are we increasingly lured into our daydreaming default mode, but our frontal lobes may fail to perform their top-down enforcement job of blocking out distractions. Shown faces and scenes and told to concentrate only on faces, older brains—for just a millisecond—let distracting and irrelevant scene information sneak in. The older brains then quickly adjusted and began to block out such distractions. But in that tiny moment the floodgates were opened and focus was lost.

This may be how a slower processing speed interferes with our memories as we age. Our frontal lobes may take too much time to tamp down interference, so we get too much neural "noise" at the start. And studies show that those who have the most initial interference also seem to have the most trouble forming solid memories or staying focused on what they are doing or saying.

"If, in the first second, you don't suppress some of the incoming information, that means you get too much information in at once and that's bad because once that information is in there, it's in there," Gazzaley explained. "With some older brains the suppressing machinery of the prefrontal cortex [part of the frontal lobes] is not coming on line fast enough and it's letting irrelevant information in."

And while most of Gazzaley's studies were done with adults past the age of sixty, there's ample evidence that such difficulties can begin much earlier, in middle age, a time when our brains can begin to be more tempted to take a rest and space out in our default modes while too much useless information rushes in. "We see this at age forty being kind of an intermediate problem," said Gazzaley.

Diverging Brain Powers

But here we have to stop, because while these difficulties arise in many brains at middle age, they do not occur in *all* brains. Nearly every study that spans ages from the forties to the early or mid-seventies—and sometimes later—shows astonishing variability. Brains are obviously varied at any age, but in middle age, that range of variability starts to increase. Some brains still operate with a knife-edged clarity, others have grown duller—most are somewhere in between. And that means that huge declines are not inevitable. As Marilyn Albert, the longtime neuroscientist at Johns Hopkins, said recently, the "true hallmark" of the brain at midlife is "variability."

"So now we have developed two categories: the age-impaired and the age-unimpaired," Albert said. "The question is, what is the explanation for that? Do those who are doing well have no age-related brain structure decline or, more likely, have they developed adaptive strategies?"

In fact, this is *the* key question. Why do some brains age well while others don't? And can we more accurately define normal aging as opposed to true pathology, such as Alzheimer's? Can we find out what makes the difference? Is it inborn or will adaptive strategies work? Over and over, scientists have been struck by the fact that it is in middle age when brains start to show not only slight declines but larger differences among one another. And it's not just human brains. While mental scores are scattered at any age, studies in a range of animals have found that the rate of variability in those scores starts to rise markedly in middle age. This is when paths begin to diverge in earnest.

"There is enormous variability and we see this variability across species," Albert said.

Indeed, a close look at one of our closest relatives—the rhesus monkey—is now confirming this, too. At a lab in Boston, an intriguing study of the middle-aged brain is still ongoing. And while it, too, is finding some downward trajectories, it shows a surprisingly wide spectrum—some doing okay, others not.

Not long ago, to see all this, I spent an afternoon at the lab of Mark Moss at Boston University School of Medicine, and, more specifically, with Bojangles, a rhesus monkey that was putting on a pretty good show with his own monkey frontal lobes when I caught up with him.

Through the years, one of the best tests of a human's frontal lobes and their ability to focus our attention has been what's called the Wisconsin Card Sorting Test, which has been around since the 1940s. The test takers, shown a group of cards, first sort them by suit—hearts, say. Then they switch and sort by number, all nines and fives, for instance. The idea is that the brain first learns the first task, then switches to another task.

In general, our brains are set up to keep doing what they've just been doing—a brain likes a good rut. So someone taking this card-sorting test is naturally tempted to keep picking hearts. To switch, the brain must *inhibit* its urge to stay in that rut and instead move over to its new mission. One of the key roles of the frontal lobes is to inhibit urges. And if our frontal-lobe inhibitory machinery is faulty, switching from sorting playing cards by suits to numbers becomes tougher. Without a strong push to stop what we've been doing, we keep doing it. We pick the hearts when we're supposed to pick nines and fives.

It's a classic test, still being used. And no one ever thought it could be used on monkeys. But it turns out monkeys do this pretty well, too. A few years ago, Moss, chairman of the neurobiology department at Boston University School of Medicine, a gregarious, out-of-the-box sort, was studying the aging brain of the monkey and decided to see if he could teach a version of the card test to monkeys. And, as he told me when I went to see him at his office, still surprised, "Lo and behold, we could."

To show how this works, Moss took me to the lab near his office to see Bojangles put on his show. Kept in a large box to limit distractions, the monkey was shown three things over and over on a computer screen: a red triangle, a blue star, and a green square.

Bojangles first had to learn that he would get rewarded only if he picked the red triangle. And through a process of trial and error, he figured it out and got an M&M. Then the game switched and Bojangles was rewarded only if he selected the blue star. The idea was to find out how long it took Bojangles to catch on and switch from red triangles to blue stars. Would his frontal lobes kick in and suppress his urge to keep picking red triangles? Could he figure out how to keep getting those M&M's?

He did. After a few false starts, Bojangles picked blue stars and got his M&M's. But there was a catch. Bojangles was a young adult at age six, which is equivalent to about age eighteen in humans. He was still a teenager. And Moss has found that this task, generally, has not been as easy for older monkeys. In fact, after studying forty-one monkeys, Moss found that difficulties clearly begin in middle age. It was the first time anyone had managed to do such a large test on the middle-aged monkey's brain. And the news was not all good.

The findings "showed that middle-aged monkeys, like those of advanced age, were significantly impaired on the conceptual set shifting task," Moss wrote in his groundbreaking 2006 study, which was published in the journal *Neurobiology of Aging*.

The test with the monkeys generally mirrored what has been suggested in studies with humans. But with humans, there was always a nagging question: Did any difficulties in middle age stem from a brewing case of preclinical Alzheimer's or vascular disease— or were they simply a part of normal aging?

Through the years, it's been notoriously difficult to tease out the difference, especially now that we know that dementia probably begins much earlier than anyone ever realized—and long before it's evident in behavior.

But monkeys don't get Alzheimer's. So, if monkeys are screened for other vascular problems, they can be a fairly good model for what happens in normal, healthy human brains as they age—and not everything, it seems, always goes right.

Continuing his research, Moss has since scanned the brains of monkeys as well as examining their brain tissue. Aging is no simple process, but Moss is convinced that one of the biggest culprits in the aging brain may be selective declines in white matter—the same white matter whose overall growth helps us to get so smart to begin with.

The brain, as we've said, is made up of gray matter—the neuron cell bodies—and white matter, the long arms of the neurons that extend throughout the brain, sending signals from one neuron to another. As we age, the arms are coated in that sheath of fat called myelin. That fatty layer allows the signal to move much faster and be timed more accurately.

Overall, myelin increases up to the fourth, fifth, or even sixth decade. But Moss and others have discovered that starting in middle age, in some people—and in some monkeys—it can also start to erode in selective areas.

In most cases during that time, such erosion is meaningless—the decreases are still outweighed by the increases and brains function better than ever. But in a few middle-aged monkeys Moss found net decreases in white matter. And he now has preliminary data showing that those monkeys that have the most negative changes in their white matter do the worst in the card-sorting games.

At this point, much is still unknown. It may be that while overall myelin increases are occurring and helping the brain operate, decreases in certain areas, such as the frontal lobes, whose rich connections need the highest efficiency for processing, may prove detrimental. It's still unclear why some brains manage to keep up repairs and some don't.

In any case, Moss was taken aback—and somewhat dismayed—to see any such problems in the brains of some middle-aged monkeys because these structural declines had not been detected before. The "findings of a marked impairment . . . in monkeys of middle age was initially of some surprise . . . very little is known about the age of onset of cognitive decline. The study demonstrated that middle-aged monkeys, as young as 12 years of age (equivalent to approximately 36 years in humans) already show impairment . . . deficits in EF [executive function] may occur earlier in the aging process than previously thought," Moss wrote rather depressingly when his first data was published in 2006.

What's more, over the past few years a clearer picture of how certain defects occur in the brains of humans has also started to emerge, and not all the news is good there, either. For this news we have no one to thank more than Naftali Raz, a neuroscientist at the Institute of Gerontology at Wayne State University in Detroit, who has become an expert in the details of brain decline and aging.

The only working neuroscientist I ran across whose papers are sprinkled with quotes from Sophocles ("For the gods alone there comes no old age, nay nor even death; but all other things are con-founded by all-mastering time," from *Oedipus at Colonus*), Raz has a succinct and downright scary way of describing what is happening in our aging brains. For instance, in a review of the scientific literature in 2006 with Karen M. Rodrigue entitled "Differential Aging of the Brain: Patterns, Cognitive Correlates and Modifiers," he wrote:

Postmortem studies of individuals within the adult age span reveal [a] panoply of age related differences in brain structure. The gross differences include reduced brain weight and volume, ventriculo-megaly and sulcal expansion. Microscopic studies document myelin pallor, loss of neuronal bodies in the neocortex, the hippocampus and the cerebellum, loss of myelinated fibers across the subcortial cerebrum, shrinkage and dysmorphology of neurons, accumulation of lipofuscin, rarefication of cerebral vasculature, reduction in synaptic density, deafferation, loss of dentritic spines, cumulative mito-chrondrial damage, reduction in DNA repair ability and failure to remove neurons with damaged nuclear DNA.

Overall, in fact, Raz has estimated that our brains shrink by about 2 percent a decade as we age. He and others talk of a "dark side to plasticity," which means that the areas of our brains that change the most during our lives, those that are the most sensitive to our environments—our valuable frontal lobes—could potentially suffer the most in the aging process.

Variability

But as dire as all these descriptions may be, Raz and Moss join with most other neuroscientists working today to stress that the main characteristic of the brain as it ages, and in particular the middle-aged brain—according to all we know now—is not universal decline but variability.

Moss, who was so surprised to find fairly significant declines in some middle-aged monkeys, was just as surprised to find out that the monkeys' brains at midlife were both "mostly all right" and highly "scattered" in terms of white matter loss as well as performance on mental tests. So what is "normal" remains up for grabs—which, again, looked at one way, is very good news.

"In the end, we had to divide the group into successful and less successful agers," Moss told me. "There clearly are these pristine agers who, even in old age, seem to be doing fine. It could be that they adapt better."

And we are beginning to find out what is most detrimental to our brains. Even if certain diseases are not yet evident, they may still, even at very early stages, be affecting them.

"There are a lot of individual differences in cortical shrinkage rates plus there are a lot of factors other than aging that cause and influence it," Raz wrote in his latest e-mail to me. "For example, vascular disease and cardiovascular risk factors such as hypertension (even relatively mild and responsive to medication) affect the prefrontal cortex and the hippocampus, both 'age sensitive' structures."

But he also is quick to add that overall, because normal levels of decline—the kind most of us experience if we're generally healthy—in middle age appear to be relatively small, for a long span of time in our modern middle age, all this might not matter all that much.

Even more important, scientists now know for certain that large numbers of neurons do not die off. The fundamental building blocks of our brains stay put. While "age-related differences in regional brain volumes and integrity of the white matter are associated with cognitive performance," Raz pointed out recently, a "review of the literature reveals that the magnitude of the observed association is modest." And, since our brains do not necessarily age in exactly the same way—and do not include a wholesale die-off of neurons—Raz sees promise, precisely because of the striking variability that he and others have found in middle age.

"Aging—a biological companion of time—spares no organ or system and in due course affects everything, from cell to thought," Raz has written. "[But] the pace of aging varies among individuals, organisms, organs and systems. And the very existence of such variability merits some measure of hope. If the positive extreme of healthy aging can be made more prevalent and if its worst and most negative expression can be delayed if not completely eliminated, the viable and enjoyable segments of the life can be prolonged into the later decades of the lifespan. In other words, successful aging enjoyed by relatively few may become the norm."

And while this may be taking things too far for some, there are even hints that, in a few instances, a little age-related brain decline—even in the area of focus—may work in your favor.

The science on this is in its infancy, but some recent studies have shown that letting some irrelevant information sneak into our brains can actually prove useful at times. If older people are asked to read passages that are interrupted with unexpected words or phrases, they read much more slowly than college students. But later, when both groups are asked questions whose answers depend on those distracting words and phrases, the older people are much better at solving the problem. As we age, we seem to be able to grasp the big picture better. But since our brains are also more easily distracted—when we are not asked to focus on underlying meanings—we let in random information that, while it can be an interference, can also at times prove handy. In these studies, the older people came up with the right answer precisely *because* they had seemingly irrelevant information stuffed somewhere in their brains.

A broader, less focused attention span, says Lynn Hasher, a neuroscientist at the University of Toronto, who is leading much of this new work, may allow a middle-aged person to know more about a situation—at times a real benefit in an often chaotic world where it's not always clear what will be pertinent in the end. Maybe there's a seemingly useless piece of information in a memo that later has meaning. Maybe while listening to another person speak, your older brain cells—wandering around here and there—notice what's happening on the sidelines. Maybe that the person is yawning or fidgeting—tidbits of information that could help you more fully evaluate that person later on.

Indeed, other studies have shown that when an older person meets someone for a second time, she already has a great deal more peripheral information, gathered unwittingly from the first meeting, than a younger person has. As Hasher says, a brain may be a little fuzzy on some details (a name, perhaps?), but may subconsciously register other information that proves to be more crucial—this person seems confident or looks shifty, for instance.

"It's not that people are doing this on purpose and saying, 'Oh, that might be relevant later on, I better pay attention to that,'" Hasher explained to me when I spoke to her about all this. "Essentially, it's like being on autopilot. It just happens. But in the everyday world, I think we overestimate the importance of deliberately doing things and we underestimate the importance of the automatic things we do. That's what keeps us from tripping, from walking into walls."

Hasher, who is sixty-three, concedes that it is not always a benefit to be on autopilot, when you have to drive on the crowded freeway or watch a toddler. But still, in more situations than we realize, this wider perspective—this lack of filtering—may help, not hurt.

"The full story is not in yet but it is amazing," said Hasher recently, adding, as she summed up the latest research, "These findings highlight the notion that cognitive aging is characterized by both losses and gains, and that whether to consider reduced inhibitory control as a help or a hindrance depends entirely on the situation."

What's more, there's also the suggestion that this less-than-straight-arrow attention can sometimes lead to art. Studies have shown that those brains that block out less tend to have more creative ideas. As Hasher says, if one part of creativity is "putting normally disassociated ideas together," then an older brain could, almost by its very nature, be more likely to come up with something quirky, new, even beautiful.

Jacqui Smith, a longtime researcher now at the University of Michigan, agrees that a tendency toward distraction can, in the right context, lead to wonderful things. "If you don't focus on one central thing, if you are thinking of all sorts of different things at once, sometimes you can come up with new associations. It's hard to measure, but this is divergent thinking; this is creative thinking. And if you're lucky, you get a true insight, something brand new."

Not along ago, after I told a friend, a poet who had just turned fifty, about the link between distraction and creativity, she just looked at me and laughed. A growing ability to daydream? No problem. Mind wandering and linking odd things together in new ways. No problem. "That's not all bad, you know," she said. "In fact, that's all very, very *good* for poetry."

7 Two Brains Are Better Than One

Especially Inside One Head

Of all the unique abilities of the middle-aged brain, perhaps none is as strange—or potentially promising—as its talent for bilateralization.

Bilateralization? Hardly a word to capture the imagination. In fact, as I was looking into this bizarre phenomenon, I kept a file labeled, a bit more enticingly, "Two Brains."

Still, that's silly, not to mention inaccurate. But there *is* something odd taking place and it involves another trick the brain learns, perhaps—though this isn't completely clear—out of a sense of panic. Sometime in middle age we begin to develop the ability, when faced with a perplexing problem, to use both sides of our brain instead of one. It's much like using two arms instead of one to pick up a heavy chair, which is not only a better way to lift a chair but may also be a more efficient way to use a brain—and part of the reason we begin to see the big—connected—picture.

Indeed, this two-fisted flair is yet more evidence of how distinctive the middle-aged brain actually is. While it is a characteristic seen later in life, too, this bilateral talent often starts in middle age and may be one of the adaptive strategies some brains adopt to stay strong. "What's really, really amazing, if I had to name a single thing," said one scientist speaking about the brain at midlife, "is bilateralization."

Admittedly, this, too, is not completely good news. Using two arms or two brain parts to accomplish what one arm or one brain part could pull off when younger signals a lack of something somewhere, compensation for a weakness or an absence, you might think.

But an intriguing aspect of this two-brain phenomenon is that it's not the weakest brains that do this but the most robust ones. A series of recent studies has found that it is the most capable who resort to this trick. It's as if the best and the brightest older brains, accustomed to being held in the highest esteem, simply refuse to give in. Without breaking a sweat, the old pro steps up to the plate and swings for the fences.

"It's nice to find out that the brain does this positive thing. It's not all passive acceptance," says Roberto Cabeza, a neuroscientist at Duke University who helped to uncover this neuro trick. "Instead, we use what we have left better. It is really encouraging. And it might have the greatest impact at middle age because we're not retired, we're still working. We may need it the most."

It's also not what scientists expected to find when they finally got tools, such as MRIs, to peer directly inside working brains. They thought they'd find the opposite. For many years it was widely believed that as the brain aged, it used much less of itself, not more. Indeed, the working model of the aging brain was something akin to brain damage. As the brain got older, most believed it became lazier. Earlier cruder measures routinely found that most brains stopped trying as hard, firing up fewer neurons. Older brains were feeble brains.

But that view has now been turned inside out, too.

Old Brain Models Disappear

"Underactivation fit well with a brain-damage model of aging," explains Patricia Reuter-Lorenz, a neuroscientist at the University of Michigan, who recently wrote a summary of this new view in the journal *Current Opinion in Neurobiology*. "The largely unanticipated result from functional neuro-imaging is *overactivation* or greater brain activity in older . . . adults."

Cheryl Grady, the neuroscientist at the University of Toronto, was one of the first scientists to get a peek at this. In the early 1990s, when Grady was at the National Institutes of Health, she became intrigued by the idea of watching the aging brain with the first real machine that could—a positron emission tomography (PET) scan. A PET scan measures changes in blood flow as brain regions activate, and Grady wanted to find out if an older brain acted in the same way as a younger brain in routine tasks such as matching faces. She began with a set of well-accepted assumptions: Older people would be much worse at this than younger people, and they'd muster fewer brain cells for the task.

Instead, she found neither premise to be true. To her surprise, the older adults performed just as well as the younger ones, and they consistently used more of their brains, not less. They tapped into the same brain circuits as the younger adults, pathways known to be active with face matching. But they also recruited an additional region—their powerful frontal cortex. This was not a matter of brains being distracted and falling into their unhelpful default daydreaming modes. Rather, the most functional brains grabbed their most powerful tool—getting the additional boost they needed from their frontal lobes.

"This was a surprise. We just wanted to see if the older people used the same pathways as the younger ones when they matched faces," said Grady when I spoke with her not long ago. "We expected that they would do worse on the task and the working model was that the older people would have less activity in their brains. And we certainly did not expect to see more frontal activity. We thought, what the heck does this mean? It was amazing and the whole field has been chasing it since."

Grady herself has led much of the chase. Just a few years later, in 1997, Grady, along with Robert Cabeza, wanted to determine whether this was simply a fluke. Was it just something that older brains did with relatively simple jobs or would it show up when tackling harder problems that, at any age require help from the elite frontal lobes. To find out, the researchers scanned the brains of young and old adults as they tried not only to learn pairs of words—*parents* and *piano*—but also to recall the correct match later on, a complex job in the brain.

And there it was again. Younger brains, as expected, used only the left side of their frontal lobes to first learn the words—called encoding—and switched to the right side of their frontal lobes to retrieve that memory, a pattern that had been well established. They use only one side of their brain at a time for this complex task.

But older adults didn't fit the pattern at all. They used their brains in a new way. They not only engaged less of their frontal lobes' left side to form the word memory initially, but they then proceeded to use *both* sides, right and left, to do the harder job of recalling the words.

This was a classic case of what many began to call bilateralization, two sides doing the work once done by one. And scientists began to find that it not only occurred in brains as they aged, but that it was the higher-functioning brains that did it. Faced with a challenge, they tapped into whatever they had, to do what they needed to do.

This flies in the face of the idea that it's better for an older brain to act exactly as it had when younger. Maybe that's not always the case. "Performance is better if the brain uses two sides," says Cabeza. "As the [older] brain reorganizes its function, it adds neural possibilities."

Older brains use more brainpower, more neural juice, to get the job done. And that often begins in middle age. But why? The best explanation is that brains learn to do this as they age because it works. After all, older brains are not recruiting areas willy-nilly. Rather, they call primarily on the part of the brain that helps the most, the frontal lobes, the region that can, as Grady says, "help you perform."

In all likelihood, this does not come without a downside. If so much brain real estate is being devoted to one thing we're trying to do, we can expect to come up a little shy when we try to do something else at the same time. Multitasking taxes the brain at any age (think teenagers texting while driving), but we often get progressively worse at it as we age.

"Overactivation . . . might have a hidden cost. To the extent that older brains engage more neural circuitry . . . [they] are more likely to reach a limit on the resources that can be brought to bear," Reuter-Lorenz has warned.

Better Brains Learn the Trick

But if, as the latest studies indicate, it's only the savviest brains that learn this trick and do it in the savviest manner—it seems a sign of calculation, not capitulation.

In one 2002 study by Cabeza, for instance, this two-brain phenomenon was distinctly linked to higher abilities overall. A group of older adults was divided by high and low abilities—all within the normal range of cognitive skills. Then they, along with a group of healthy younger adults, were given relatively complex tasks, in this case—again—word-pair matching. As expected, the younger adults, scanned by PET, used the right sides of their brains and did fine on tests of mental skills. The older adults who used only their brains' right sides, however, were also those who had scored on the low end of cognitive ability. They used the same brain area as those who were younger but used it less efficiently.

But the older adults who used *both* sides of their frontal lobes were the cognitive champions. Indeed, the pattern was so recognizable that Cabeza, in the 2002 study "Aging Gracefully: Compensatory Brain Activity in High-Performing Older Adults," gave the pattern a name—HAROLD, for Hemispheric Asymmetry Reduction in Older Adults. Translated, that means that if we're smart, we figure out how to recruit as much brainpower as we need.

Over the last several years, this idea has been pushed even further, with added twists. Cheryl Grady and psychologist Mellanie Springer at the University of Toronto, for instance, recently found that younger adults used mostly their lower-level temporal lobes to solve a certain memory problem, but older people who performed well instead used their higher-level frontal lobes. And even more interesting, it was those adults with the most education who tapped into this premier brain region.

Is it possible that those with more education had, through the years, simply grown accustomed to drumming up high-level brain reinforcements when necessary? Maybe so. As Grady herself concludes: "The higher the education, the more likely the older adult is to recruit frontal regions, resulting in better memory performance." Higher education seems to enable older adults to "call up the reserves."

This suggests that those of us who learn to call up more of our brains are better off in the long run and that this brain trick, as Grady says, has "some functional significance. Older adults who are better able to recruit more areas are the high performers," she says. "It means in some cases, as we age, it is not just a matter of the brain turning down but the brain turning up."

It may be, too, that we don't bother to use two brain sides or recruit higher brain areas when we're young because with a relatively new brain, it's just not necessary. With an increase in "neural noise"—that interference from irrelevant information—however, the brain turns to its most powerful region to help it focus.

"It's like shouting in a quiet place, which doesn't make any sense and is not an efficient way to communicate," explains Cabeza of Duke. "But in a noisy place, shouting can work better and is a more effective means of communication."

Building a Brain Scaffold

Recently, Patricia Reuter-Lorenz and Denise Park of the Center for Brain Health at the University of Texas at Dallas have rolled all these insights into one tidy concept called "scaffolding," which makes the case that brains are set up on purpose to constantly reorganize and recruit more brain tissue as needed. Our brains are built to roll with the punches, and better—or more carefully cared-for—brains roll best.

"What we think is happening," Park told me, "is that the brain is continually building new scaffolding, responding to the changes and tiny insults by attempting to rewire and reorganize itself."

And, Park said, "I suspect that middle age is a kind of crossroads for all this, when the brain either learns or does not learn these new patterns.

"If we maintain good brain health, we build better scaffolding and our capacity to adapt continues. That means that if you make good or bad decisions or good or bad events occur, then that will have consequences. It's just like if you wear out your joints in your thirties, it might not bother you until you are seventy. It's the same with the brain. In middle age, it matters what you do."

Or as Patricia Reuter-Lorenz sums up, quite encouragingly: There's recently been a "shift from the dismal characterization of aging as an inevitable process of brain damage and decline. Instead, the emerging story . . . is that aging can be successful, associated with gains and losses. It is not necessarily a unidirectional process but rather a complex phenomenon characterized by reorganization, optimization, and enduring functional plasticity that can enable the maintenance of a productive and happy life."

Bilateral Sparkle

Indeed, there are those who go even further and suggest that this bilateral use of brainpower may be *the* key ingredient in the power and creativity of our middle-aged brains. It's been shown, for instance, that some bilingual older adults, having developed the ability to flexibly negotiate two languages in separate parts of their brains throughout their lives, have smaller age-related declines in brain function. That suggests that those who establish such patterns of using more of their brains early on may be in better shape along the way in a range of different ways.

Gene Cohen, a longtime researcher of aging and author of *The Mature Mind,* who has studied the connection between art and neurons, thinks creative thoughts and solutions can also be partly traced to this ingenious brain trick. As we age, the two sides of our brains become more intertwined, letting us see bigger patterns, have bigger thoughts—reaching, he believes, the level of art.

"The brains' left and right hemispheres become better integrated during middle age, making way for greater creativity. . . . The neurons themselves may lose some processing speed with age, but they become ever more richly intertwined . . . that's why age is such an advantage in fields like editing, law, medicine and coaching and management," Cohen has written.

"As our brains become more densely wired, they also become less rigidly bifurcated. In most people the left hemisphere specializes in speech, language and logical reasoning while the right hemisphere handles more intuitive tasks such as face recognition and reading emotional cues . . . but this pattern changes as we age. . . . Older [people] tend to use both hemispheres. . . . This neural integration makes it easier to reconcile our thoughts with our feelings."

Robert Cabeza does not think this idea is crazy at all.

"Maybe that means that if you are doing a task cross-hemisphere you will simply do better," Cabeza suggests. "We see it with physical things . . . moving a chair with two hands or bending your knees to pick something up. That may be a better way to do things altogether; it prevents injury and it's better for the whole body. If wisdom is learning to use the brain in different ways, well, maybe, in the end, that works better, too."

The Genetic Road Map

In the backroom of a lab at Harvard Medical School sits a large box freezer whose contents are kept at 140 degrees below 0. If you lift up the lid, you see rows of small plastic containers, each filled with tiny bits of human brains. The brain samples, carefully coded and cataloged, are from adults ranging in age from 26 to 103.

And it's in those brains, deep in their microscopic folds, that scientists have found even more evidence that brains embark on different journeys in earnest at midlife. And while those divergent paths may relate to levels of education and other adaptive strategies, the brain's course is also determined by our genes as well.

We are born with a genetic road map, with certain genes—or segments of the DNA in our cells—programmed to activate certain proteins that, in turn, regulate how our bodies and brains function. But those genes—that DNA—can be damaged along the way, perhaps through a preordained process of normal aging, by the environment we live in, such as the level of toxins we are exposed to, and by how *we* behave in that environment—whether we exercise, eat right, hit our heads, or get a disease. When damage occurs, genes might not be able to properly tell the proteins in our brain cells what to do.

The scientists tending the frozen brain cells wanted to see if they could find a pattern of gene activity in brains at various ages. And they did. After scanning brain samples with a gene chip, a newly invented tool that can measure gene damage and activity, Harvard researchers found that in terms of overall condition, brains of those under forty and past seventy-three years of age look pretty much as you'd expect: little damage and lots of activity in the first group, and more damage and less action in the other.

But the middle-aged brains were all over the map. The brain of one forty-five-year-old man most resembled that of an average brain of a person over seventy. And the brain of a fifty-three-year-old woman matched those of people in their thirties.

"Individuals may diverge in their rates of aging as they transit through middle age, approaching the state of 'old age' at different rates," said Bruce Yankner, the Harvard neuroscientist who examined the brains in the freezer. Though the brain samples were obtained from cadavers or those undergoing brain surgery, and those conditions could have had some impact, overall the brains were considered normal and indicative of what would be found in healthy, living adults.

"At middle age," Yankner said, "brains become particularly variable."

The study by Yankner and his colleagues, published in 2004 in the scientific journal *Nature*, was the first to take a systematic look at how a brain ages on its most basic level, its genes, using these gene chips or microarrays.

Yankner's studies, more than any others, have taken advantage of this new technology to try to figure out what is happening at a genetic level as the brain ages. His team has now looked at twenty thousand protein-linked genes in the brain and they've found the very same thing twice.

The changes in genes that are affected by aging (about 4 percent of the total) generally begin in our late thirties, much earlier than expected. In particular, Yankner found age-related changes in about twenty genes that are crucial for learning and memory and brain-cell flexibility.

But there's good news here, too. Yankner found that around the same time—late thirties to early forties—another group of nurse genes steps in to help out. These are the genes that protect and repair neurons from damage, and they begin to work overtime, perhaps delaying the net impact of the damage. This, too, could be part of the reason why cognitive decline often doesn't show up until later in life—and why some retain their intellectual prowess longer than others. Perhaps some people simply have more nurse genes—or for some reason have been able to retain better-functioning nurse genes—than others. And these microscopic differences begin to show up in middle age.

"There seems to be a similar profile for the young before age 40 and a similar profile after age 73. But the most variable group was between ages 40 and 70. Right around middle age you can see the transition in age-related genes. They were just not aging at the same rate. Some resembled the young and some were more like the old. It was very striking.

"The research is still in early stages, but these changes in genes for synaptic function, learning, and memory could help explain the subtle declines in middle age in abilities such as short-term recall," Yankner told me when I spoke with him recently.

But, he added, more optimistically, other genes are stepping up to the plate. "These are the ones that protect the cell from damage, help build new connections. So at the same time that there is this decline, there is this compensatory activity kicking in, too. I would guess that with most at middle age, there's probably a balance between the two."

Of course which side wins that balance game is obviously crucial. Yankner likes to talk of one ninety-three-year-old woman who was in nearly perfect cognitive shape when she died and donated her brain to Yankner's lab. And perhaps not surprisingly, when the researchers took a look at her brain, they found she had the genetic brain patterns of a middle-aged person.

"We know that she was cognitively intact and her brain was, too," Yankner told me.

How can some get so lucky? Is it something they ate or read or did? And how can some, on the other end of the spectrum, get so unlucky, brain-wise? Are the lucky ones better from the get-go, or, as Yankner suspects, do they have—or develop—better repair mechanisms and adaptive strategies in their brains? And do these mechanisms mean that they remain able to call on more of their brains—two brains—to help them out?

It's possible that all this is simply following some set genetic program. Perhaps along the way, Yankner speculates, evolution produced choices. It may have proven over time that it was more important to keep our heart muscles going strong and let our brains, in particular short-term memory, slip a bit. After all, it might be more important to get our hearts pumping to get away from an angry tiger than to recall exactly what we ate for breakfast.

Still, he, along with most, believes it's likely that in the end we will find that changes in the aging brain arise not just from genes alone but from a combination of our DNA and the soup it lives in— our environment and the way we live our lives. And that means we can make a real difference—and what we do for ourselves during middle age may be particularly crucial.

"There are a lot of redundant systems built into human cells to repair damage. There is a good system to keep the brain intact," said Yankner, who at age fifty is right at the brain crossroads himself.

"What's more surprising is that this system breaks down at all. That is the great mystery of aging."

8 Extra Brainpower

A Reservoir to Tap When Needed

So if our middle-aged brains are—on balance—so masterful and marvelous, what can we do to keep them that way?

For an answer to that, there's no better place to start than with a now-dead nun.

Her name—or the name she's been given in the scientific literature—is Sister Bernadette, and she's given us provocative clues to what may be the brain's most powerful ploy. Sister Bernadette was a part of what's been famously called the Nun Study. Since 1986, University of Kentucky scientist David Snowdon and his colleagues have studied 678 Catholic nuns in an extraordinary experiment to look at how the brain ages and why.

As part of the study, the nuns, members of the convent of the School Sisters of Notre Dame, have had periodic mental tests—how many animals they can name in a minute, how many coins they can count correctly, how many words they can remember after seeing them on flashcards. Through the years, they've also provided personal information—who their parents were, what illnesses they suffered, how many years of schooling they had—details that have been meticulously cataloged and stored in convent archives.

And perhaps most important, the nuns all agreed that after they died they would donate their brains, which are placed in plastic tubs and shipped to a laboratory where they are stored and analyzed.

Nuns are a particularly good study sample because you can usually eliminate a slew of activities—heavy smoking and drinking, for instance—that aren't particularly good for a brain and skew results. And the Nun Study has had a score of fascinating findings, including suggestions that dementia may be linked to small strokes or insufficient folic acid in the diet. And, in an especially striking result, the nuns who used the most elaborate sentences—packed with more complex and optimistic ideas when writing autobiographies in their twenties—had a lower risk of dementia decades later.

Still, in the midst of such revelations, the story of Sister Bernadette stands out. Among the nuns, she was a class star. Early in her life, she had earned a master's degree and taught elementary school for twenty-one years and high school for another seven. At ages eighty-one, eighty-three, and eighty-four, she aced any cognitive test thrown at her. Then, about ten years ago, after she died of a massive heart attack at age eighty-five, Sister Bernadette's brain was sent to be analyzed, initially without its identity being known.

At first glance, the brain seemed fine. It weighed 1,020 grams, about normal. But, as Dr. Snowdon writes in his moving and personal book, *Aging with Grace*, about his relationship with the nuns he both studied and grew to love, a microscopic look at Sister Bernadette's brain soon revealed something far different. There was, Dr. Snowdon writes, "little doubt that Alzheimer's disease had spread far and wide. Tangles cluttered her hippocampus and her neocortex all the way up to the frontal lobe. Her neocortex had an abundance of plaques as well." In fact, on one scale used to determine the degree of Alzheimer's, Sister Bernadette rated the most severe, Level 6.

How could it be? How could a woman who was, up until the moment of death, a cognitive champ also have extensive plaques and tangles, the hallmark of Alzheimer's disease? Expressing his own surprise at the time, Dr. Snowdon said, "Despite an abundance of plaques and tangles in her neocortex, the function of that brain region seemed to be incredibly preserved. It was as if her neocortex was resistant to destruction for some reason. Sister Bernadette appears to have been what we, and others, have come to call an 'escapee.' "

An escapee?

Is that possible? Was there something in Sister Bernadette's background—her richly endowed sentences, perhaps—that, although her brain had all the physical signs of dementia, had somehow protected her?

On one level it would be easy to dismiss the case of Sister Bernadette as interesting but odd. It would be easy if she were a fluke.

But she is not.

Take the case of the retired professor from London. The professor—he's called the Chess Player in scientific studies—loved to play chess and was uncommonly good at it. As he played, he could easily think seven moves ahead. But at a certain point, he noticed a change. Although his wife and family thought he was fine, he was worried. He found he could think only *four* moves ahead. Convinced there was something terribly wrong, he went to the clinic of Nick Fox, a neurologist at University College London's Institute of Neurology. No problems were found. The professor cruised through a battery of tests intended to detect early signs of dementia. A brain scan was normal. The professor, then seventy-three, continued to play chess, read history books, cook elaborate meals, do the family's finances, and even learned how to use a computer. He also continued to have brain scans, which detected few significant changes.

Then a few years later, the professor died of causes unrelated to his brain. And much to the surprise of Fox and the professor's family, an autopsy showed that the Chess Player's brain, too, was riddled with the plaques and tangles of Alzheimer's. The professor had what appeared to be an advanced case of dementia. Yet for years the only outward sign of it was that he could think four chess moves ahead instead of seven.

How could that be? How could a brain so ravaged by disease still be functioning at such a high level? Had something shielded the brain of the chess-playing professor? Was he, like Sister Bernadette, an escapee?

Who Escapes and How?

For many years, scientists have puzzled over why some people seem to withstand brain injury better than others or why two people can have strokes of the same severity and yet one suffers severe impairment and the other recovers.

The differences have been particularly perplexing to neuroscientists because most believed that healthy brains, aside from a few IQ points here and there, were pretty much the same. After about age three, according to long-held scientific thinking, a window of opportunity began to close. Sure, we could polish our French, but the basic structure of the brain was thought to be largely set. And in many ways, that view made sense. Unlike other cells in the body, brain cells don't divide, so the same neurons stick with us for as long we stick around. With age, some brain cells die off, but it was thought that they were not replaced, leaving our brains to accumulate all the junk and insults thrown their way. Indeed, major changes were considered not only impossible but, if they did take place, mostly bad.

But that view has changed now, too. Even the most conservative of neuroscientists agree that brains can be tinkered with, perhaps even vastly improved, on their most basic synaptic level throughout our lives.

And the very fabric of our daily lives—how we spend our workdays and even our vacations—may influence how we respond to disease, brain injury, or even the more nuanced shifts that come with age.

This is the fundamental idea behind what's now called "cognitive reserve," that some brains have—or can develop—a reservoir of strength that, when the going gets rough, offers protection, perhaps much like the brains of Sister Bernadette or the Chess Player. It's not that those with cognitive reserve are smarter in the conventional sense. Rather, they seem to have an emergency stash of brainpower—perhaps stronger, more resilient, or more efficient brain connections or repair systems that can be called up when necessary. Certain brains may develop a sort of mental padding that allows them to tolerate more damage.

"When we did the autopsy [on the Chess Player] it was amazing that with such a relatively mild level of apparent dysfunction, he had such widespread changes [in his brain]," said Fox when I spoke with him about his own study of the Chess Player.

But if it exists, what exactly is this reserve? Can you see it? Touch it? Can you, if you want to, get more of it?

The story of cognitive reserve is still being written. It is also one of the most encouraging stories ever to be told about the brain— certainly the best news yet for the middle-aged brain. And it's a story that began not that long ago.

In the early 1980s, Robert Katzman was living in Rye, New York, working as the chairman of neurology at Albert Einstein College of Medicine in New York. In his job, he saw hundreds of patients suffering from dementia, but there was little he could do. At that time— as now—very little was known about Alzheimer's disease.

So Katzman decided to dig deeper. Scientists knew that those who died with Alzheimer's usually had plaques and tangles of fibers in their brains, but the relationship was murky. Did more tangles equal more disease? Hoping to clarify at least this aspect, Katzman began a study of a group of elderly people living in a Manhattan nursing home. His initial aim was simply to replicate an earlier finding suggesting that the level of plaques and tangles determined the severity of the dementia.

And, in fact, Dr. Katzman found exactly that. In his 1988 study of 137 nursing home residents whose brains he dissected after their deaths, he saw a clear relationship between the number of tangles and mental decline.

But Dr. Katzman also found something else. In what he called Group A, there were some patients who did not fit the pattern at all. The ten people in this group all had brains full of tangles. But, as was discovered later with people like Sister Bernadette and the Chess Player, they also had been functioning at a high cognitive level up until the very end. Lots of tangles, but mentally first-rate.

In the New York nursing home, Dr. Katzman had found the first documented set of escapees.

When I spoke with Dr. Katzman he was eighty-two years old and long retired, a professor emeritus of neuroscience at the University of California at San Diego. Still mentally alert, he easily recalled his reaction to his own escapee finding. "It was a surprise; we only set out to replicate the other studies," he said. "And then we found this. It was new, very new." (Dr. Katzman died in September 2008.)

As he wrote in February 1988:

Our study does provide remarkable findings in regard to Group A, subjects with preserved mental status but definite histological changes of the Alzheimer type. These nondemented subjects with Alzheimer's changes were functionally and cognitively as intact as those in the control group, the nondemented subjects who were free of histological markers or brain pathology.... It can be concluded therefore that there is a group of elderly with preserved mental status and Alzheimer changes.

Knowing that most new and unconventional ideas do not always find the warmest of receptions, I asked Dr. Katzman if he recalled the response of other scientists at the time. "Was it controversial? Oh, yes," he said, laughing. "I mean, it was new, so automatically it was controversial. But I know I believed it then and I believe it now," he added bluntly. "We had the data."

As it happened, he also had something more. Katzman found that the Group A brains were not only somehow protected but also bigger. As he wrote at the time:

In regard to the number of large neurons in the three regions of the cortex measured, these nursing-home residents surpassed the subjects in the control group as well as the demented patients with Alzheimer's disease . . . the brain weights in Group A were greater than in the other groups, suggesting that there has been less atrophy than normally found in the very elderly or that this group of patients started with more neurons and larger brains and thus had greater reserve.

Wondering what it all might mean, he went on:

This implies that patients in Group A had incipient Alzheimer's Disease but did not show it clinically because of this greater reserve. . . . [Those who have] retained intact pyramidal neurons and whose brains are heavier than age-matched normal subjects . . . these people may have escaped the shrinkage of large neurons that accompanies normal aging and the loss of large neurons that usually occurs in Alzheimer's Disease so mental status is preserved in spite of beginning Alzheimer changes. Alternately these people might have started with a larger brain and more large neurons and thus might be said to have had a greater reserve.

With that, the idea of cognitive reserve was officially born. It was also linked, from the start, to bigger brains. Indeed, a few years later, similar observations were made by a technician helping to dissect Sister Bernadette's brain: "Look at the initial MRI scan," the technician said. "It shows an unusual amount of gray matter." As Dr. Snowdon elaborated in his book: "As it turns out Sister Bernadette had more gray matter . . . than 90 percent of the other sisters studied."

The Education Connection

So it's possible to be an escapee, to both have tangled brains and still teach a class of ninth graders? But to do that, do you also have to have a giant brain? For those of us with smallish heads, this is a less than happy thought.

Luckily, science didn't stop there. As the research into cognitive reserve has matured, it has become increasingly (and happily) apparent that there is more to all this than buff and brawn. While there is a correlation between brain size and reserve, cognitive reserve turns out to be much more complex—and possibly within reach—than that.

Indeed, one of the most prominent producers of this extra brainpower turns out to be something that would make your first-grade teacher proud: education. Just as those with more education seem to be better able to call on more parts of their brains when needed, education also seems to offer a kind of overall protection, at least against the outward manifestations of disease.

"Education changes the brain; that is now clear," said Dr. Katzman when I spoke with him about where we are now with cognitive reserve. "I don't think we know exactly how, but it changes the brain."

In recent years, studies have found an indelible line between education levels, or, in the case of those with no access to formal education, literacy levels—and how well the brain ages. This is not a thought that comes completely out of the blue. For many years, education has been tied to living longer in general. The reasons are still being debated, but the idea remains steadfast and serious. As my colleague Gina Kolata wrote in a newspaper series on aging recently, "The one social factor that researchers agree is consistently linked to longer lives in every country where it has been studied is education. It is more important than race; it obliterates any effects of income," adding that education may "somehow teach people to delay gratification," a habit that might mean you giving up that cookie or cigarette and instead taking a walk.

Still, tying education levels specifically to brain aging has been more controversial and much more complicated. Some of the initial evidence came from the Nun Study, whose first results showed that those who had higher education levels aged more independently, and were able to bathe, eat, and dress themselves considerably longer than their counterparts with less education.

That tantalizing finding was small, however, and even David Snowdon of the Nun Study concedes that when he presented it at a scientific conference in 1988 the reaction was far from overwhelming, saying he'd had "a better audience for my 4-H project on chickens at the San Bernardino County Fair."

Then Katzman found it, too. In one of the first large epidemiological studies in China, among five thousand people living in Shanghai in the late 1980s, Katzman found that those with no education had twice the risk of developing dementia than those who had attended middle school or even elementary school. Similar results were later found in population surveys of dementia rates in France, Italy, Sweden, and Israel.

"Alzheimer's disease is a democratic process," Katzman wrote, summing up his findings later. "Physicians and psychologists, chess masters and physicists, mathematicians and musicians may become victims of this disorder. . . . Yet a number of recent community studies report that individuals with a lack of education or low education are more likely to develop dementia and Alzheimer's (AD). This . . . has profound social and biological as well as medical implications."

Reserve and Respectability

Despite such findings, however, cognitive reserve has been battling an uphill fight for respectability. Doubts and disagreements persist. No one has suggested—or ever found—that education in any way prevents you from getting plaques and tangles or becoming demented. Going to school or becoming self-educated does not eliminate pathology. But beyond that idea, there has been consensus on little else.

And the whole concept, from the start, has had a serious chicken-and-egg problem. Was it simply that those with better brains sought more education or read more and then developed even bigger and better brains as a result? Were those who were inclined to become more educated simply the people who had better nutrition while young or had lives that were generally more protected from toxins that could harm the brain?

Many did not swallow the idea of brain reserve at all. One of these people was Yaakov Stern, a neuroscientist at Columbia College of Physicians and Surgeons. Fresh out of graduate school and working in Manhattan in the late 1980s, Stern had heard stories about cognitive reserve, or backup brainpower, an idea, he says, that was "being bandied about." But he, like many others at the time, thought it was simply a matter of diagnosis. He believed that those who were more educated were simply better at doing the cognitive tests and were, therefore, less likely to be diagnosed as demented. "I thought it was diagnosis bias," Stern told me.

As his career progressed, Stern found himself with a good position and adequate funding and he decided to take a serious look at the idea of cognitive reserve.

Interestingly, he, too, focused on a group of elderly living in Manhattan. His group was not living in nursing homes, however, and had a wide range of education levels, varied occupations, and was ethnically diverse as well. Because of his concerns about diagnosis, Stern was as careful as he could be to make sure that although the participants had varying levels of education, they were nevertheless at the same level of cognitive abilities when the study began. He also screened for signs of tiny strokes or vascular problems. Then he followed the group for four years to see what would happen.

And there it was again. Stern and his colleagues found that the better educated in the group were much less likely to show outward signs of dementia. To add a new wrinkle, he also found that those with more complex occupations, which usually meant dealing with human beings rather than working with repetitive machines such as on an assembly line, were also much less likely to become demented. In 1994, he published his study in the prominent *Journal of the American Medical Association*.

"We found that those with less than eight years of education were twice as likely to become demented and those who had lower education and lower-level occupations were three times as likely," Stern said.

The finding was, again, surprising, in particular to the skeptical Stern. What was important was that in this study all the participants were at the same place to begin with in terms of mental abilities and general health. The only way they differed was in their level of education. It was also important because it followed the group going forward, with no idea what the outcome would be. Indeed, the results were so striking that Stern went full-tilt to the other side and is now a true champion of the idea of cognitive reserve.

To fully convince himself, though, Stern set out to examine this reserve from every angle he could think of. He and his colleagues found that among those with the same outward signs of dementia, those who were better educated also had the lowest levels of cerebral blood flow, a sign of a higher level of pathology. In other words, again, the more highly educated had a worse physical condition inside their brains, but something was shielding them from the full force of their dementia. Like the nun and the Chess Player, something was helping them tolerate the would-be effects of the disease better.

In two other revealing studies, Stern and his team found that demented patients who had higher levels of education or occupation declined and died faster after being diagnosed. While on the surface that seems counterintuitive, it fits perfectly with the theory of cognitive reserve. It suggests that those who can call on more brainpower can hold back the outward signs of the disease. Then, by the time the disease becomes outwardly evident, its effects are much further along in the brain and those patients both get worse and die faster. These people, more escapees, as Stern says, "have less time to live with the effects of the disease and that seems like a good thing." Better to be an eighty-four-year-old woman who has had no apparent problems and declines quickly and dies than one who spends years with feeble abilities.

Still, many researchers remained unconvinced and confused. After his first study was published, Stern got a call from a woman whose husband, a Nobel Prize winner, was suffering from a terrible case of Alzheimer's. "She said, 'What the hell are you talking about,'" said Stern.

Even though no one was suggesting—or says now—that education guarantees protection from dementia, the idea that something as amorphous as education could buffer the brain from the actual physical assault of a serious illness was a hard sell.

"We just didn't think the brain worked that way," said Stern.

But increasingly, it seems it does. Other solid studies, such as the Rush Religious Orders Study (more nuns, as well as priests) in 2004, also revealed that for a given level of severity, the more educated, on autopsy, had more tangles and plaques. Again, this suggested that those who had more education were protected longer from the most severe impact of the disease.

That same group found, too, that it was not just the level of education that was connected to the risk of dementia but also "cognitive-stimulating activities." In a landmark study that helped boost the crossword puzzle industry, the Rush group found that over a five-year period, those who had done more to activate their neurons were about half as likely to develop Alzheimer's. Stimulating activities were defined as those in which "seeking or processing information" was central and something you could do by yourself (to factor out the impact on socialization, which can also be hugely beneficial to the brain). That meant playing bridge was out, but reading magazines or newspapers, going to the library, doing word games, taking music lessons, or learning a foreign language were all counted.

A researcher in Stern's lab at Columbia also added more solid evidence. In a study of 1,772 nondemented adults, Nick Scarmeas found that even those with a higher level of "leisure activities," including walking, visiting with friends, or reading, had a 38 percent lower risk of developing dementia than those who did those things less often. And the risk of dementia decreased by 12 percent for *each* additional activity added, a finding that held up no matter what level of occupation or education.

Of course, it's still very much possible that those who fill their lives with concerts and Chinese lessons might simply be better off, brain-wise, to begin with. But the most rigorous studies have done all they can to ensure that all participants are at the same general cognitive level from the start. It's true, too, that some who do poorly may have early, undetectable stirrings of dementia or vascular disease. But as Stern's colleague Nick Scarmeas sums up in a book on the latest research on cognitive reserve, edited by Stern: "Overall, the accumulated data seem to make a case for a protective effect of physical, intellectual and social activities for cognitive decline and dementia."

Most of the recent research has looked at the relationship of cognitive reserve to dementia because, as a slowly progressing disease, dementia is easier to measure. Still more controversial has been the idea, as mentioned by Scarmeas and many others, that this mental cushioning could also soften the less forceful assaults of normal aging.

But in just the last few years there has been a shift. Many now believe this extra brain reserve is real. And it appears that the way we live our lives can have a very real impact on the overall power, strength, and staying power of our brains.

"Cognitive reserve is a very powerful idea," says Stern. "But it is also a simple one. The fact is that there is not a linear relationship between pathology in the brain and clinical manifestation of that. Something is mediating and some do better than others. And this is not specific to even aging or Alzheimer's. Those are good places to see it. The evidence is out there."

As the research into cognitive reserve has exploded, its effects can be seen in wider and wider areas. In particular, researchers now want to know what it means to people who are hitting middle age and beyond. Do we have to start building this backup brainpower as babies? In the womb? Or can we still grab a little if we tackle it when we are past fifty? How about sixty?

The most recent studies suggest that brain reserve can be built anytime in our lives. One long-term study found that those with high socioeconomic status and fully engaged in their environment had the least intellectual decline during a fourteen-year period (widowed women who had never been in the workforce and who had a disengaged or lonely lifestyle did the worst). Another study included in the Scarmeas and Stern review—of World War II veterans tested twice over forty years—found that participating in intellectual activities was related to intellectual performance later in life. Similarly, a well-known British study—a long-term look at people in England, Scotland, and Wales born right after World War II—found that social class, occupation, and education at age twenty-six helped shape cognitive ability at age fifty-three.

More recently, in 2007, a study of workers at a lead-smelting plant found that among adult men with the same blood-lead levels, a result of exposure to heavy metals known to cause neural damage, those with the highest reading scores, while not protected from declines in hand-eye coordination, were somehow shielded in cognitive areas. The men in the group of better readers performed 2.5 times as well on tests of memory, attention, and concentration tasks not necessarily related to reading. The study's author, Margit L. Bleecker, a neurologist at the Center for Occupational and Environmental Neurology in Baltimore, said she now is convinced that "the brain is like a muscle" and can be pumped up at any age. "Those who are cognitively more active, exercise more, and are more socially connected have more cognitive reserve," Bleecker says.

Certainly early development of the brain is key and head injuries along the way don't help, but research increasingly finds that reserve can be added at middle age and beyond. In a Scottish mental-health survey, children born in 1921 first had their IQ tested at age eleven and then at age eighty. While IQ at age eleven was a decent predictor of how well a person would do later on—and IQ is partially inherited—there was clear evidence that we're not necessarily stuck with what we are born with. Some in that group were able to push their scores up significantly. Something was changing their brains for the better—even well past childhood.

Granted, there are still disagreements and doubts out there. Even Nick Fox in London, who reported on the case of the Chess Player, believes that cognitive reserve has become such a hot topic that all sorts of grand and largely unsubstantiated claims are now being made in its name. Even with the Chess Player, he says, the question remains: Did he withstand the assault of Alzheimer's for so long because of "something he did" in his life—that is, read, play chess, or become highly educated—or because that's just how his "high functioning brain was"?

But most evidence now suggests that we can make a difference by what we do; we can boost our reserve, even when older.

"You are not just born with cognitive reserve, and that is the most encouraging piece of this," Stern said. "It seems to be malleable even in later life. The thing is, we are not sure what is the most protective thing you can do—is it gardening or particle physics? We need to figure that out."

Perhaps more than anyone else, Stern remains on the trail of cognitive reserve. Step by complicated step, he is trying to find out what it is in our brains that can help—and what we can do to help our brains. Some of Stern's most recent research, for instance, has confirmed that it is not just complex occupations that make a difference to dementia but occupations that are highly physical.

"I used to have this ivory tower view of cognitive reserve, that it was linked to intellectually stimulating things," Stern says. "Now I am trying to exercise, to go on the treadmill more."

Many cognitive reserve studies have looked at large populations, so-called epidemiological studies that seek out correlations or trends. But in the past few years, researchers have also taken a more focused look inside the brain. A study in 2001 by Lawrence Whalley at the University of Aberdeen in Scotland found that among those with the same outward behavioral symptoms of dementia, the most educated also had more decay in their brain's white matter, that crucial outer coating of brain cells. Again, this was an indication that those with more education were somehow able to cope with more damage and still function.

In one unusual study, researchers found that among a group of people who were depressed and received electric shock therapy known to cause cognitive problems, those who had higher levels of education recovered much faster. And a recent study by Shelli Kesler at Stanford University School of Medicine found that the more educated and those with larger brain volume even had a smaller dip in IQ after traumatic brain injury.

"I'm quite passionate about cognitive reserve," Kesler told me. "I am definitely a believer. It's just like an athlete is better at sports and would be more protected from heart disease than someone who is obese."

Experience Changes Structure

"Cognitive reserve," Kesler said, "is basically a type of neuroplasticity—we know from repeated animal and human studies that experience can alter our brain function and structure. I think cognitive reserve results from a combination of heredity and life experience. If you have smart parents, you will have a higher reserve—just like the athlete model—some people are just born with greater physical prowess—I'm trying to identify particular genes that might endow people with greater cognitive reserve or increased neuroplasticity.

"But just like an athlete—genetics will only get you so far—training and practice are essential. If you engage actively in mental and physical activities, particularly those that have a graded challenge (get more difficult as you progress), you also can increase cognitive reserve. It's best to continually increase the challenge or difficulty level to keep on benefiting . . . a variety of mental activities that are new and stimulating will be the most helpful. I think this is one reason why people with higher education levels tend to have higher cognitive reserve—they have had a variety of mental stimulation and tend to seek this out."

And, she said: "The best news is that neuroplasticity exists across the life span—you're never too old to improve your brain function."

Back in New York and now a true believer, too, Yaakov Stern at Columbia University is even trying to create cognitive reserve in his lab.

Much of what matters in the brain as it ages, Stern believes, will depend not on its hardware—how big, how many brain cells, how many branches and connections—but on its software, that is, how a brain operates. He believes that brains may age better if they also have the capacity to compensate, or "switch to plan B," by using additional or alternate parts to do what they need to do.

This is a version of the two-brain idea. And it means that those who can, or who can learn how to, use more of their brains when they need to will be better off in the long run. These are the lucky escapees, who, Stern says, are "able to summon that compensatory response. They are used to engaging these networks and can do it more easily."

Another key ingredient may also be basic brain efficiency. In one of his most recent studies, Stern found that, confronted with increasingly difficult problems, those with higher IQs used a smaller percentage of their overall brainpower to get answers. It was as if that group could "accelerate," or ramp up, their brains with less effort.

Stern firmly believes that we can, by challenging our brains throughout life—perhaps by learning how to use our frontal lobes more efficiently—build up cognitive reserve. He is also trying to figure out if such brain efficiency can be taught to the middle-aged and older brain, when such training might be most beneficial.

"The best way I can explain this is to think of two swimmers," he told me. "If you take a very good swimmer and ask him to swim a lap and then you have me swim a lap, at the end I will be winded and the good swimmer won't break a sweat. He is more efficient. Then you ask us both to swim a mile and I won't be able to do it at all but he will. He is not only more efficient but he has more capacity. Then you take that good swimmer and put a ten-pound weight around his waist; how does he do then?"

The ten-pound weight is middle age, old age, and disease. And the question on the table—one that is being pursued with a passion that would make that old seeker of the Fountain of Youth, Ponce de León, proud—is, how can we, even as we forget where we parked the car, build up reserve and keep our brains swimming?

~~~ Part Three: Healthier Brains

9 Keep Moving and Keep Your Wits

Exercise Builds Brains

Kevin Bukowski is forty-seven years old, and though he'd always been a runner, he'd slacked off lately. Then, offered free gym membership as part of a research study, Bukowski got serious about exercise. He woke up at five A.M. in his home in upstate New York, got onto a bus at six, and an hour later was on a treadmill in the gym across from Columbia University's medical center in Manhattan, where he works helping to coordinate clinical trials. Every day, three or four times a week, Bukowski did the same routine: twenty minutes on the treadmill, twenty minutes of sit-ups. And what happened? At the end of five months, Bukowski was happy to find that he'd lost a few pounds, his body mass index went down, and he felt better "emotionally, physically, and spiritually. I just had more energy and I was not as tired at the end of a long day."

But most important of all, his dentate gyrus went wild. His *dentate gyrus*? Hardly seems like something to get out of bed for, whatever it is. Dentate gyrus?

Well, yes. Over the past few years, the dentate gyrus, a small section of the hippocampus, an area crucial for memory, has emerged as a superstar in the story of the brain as it ages. And the dentate gyrus, it turns out, is particularly fond of exercise. Not long ago, in fact, the dentate gyrus caused a bit of a stir at a Columbia University lab. Early one afternoon, a group of scientists was watching a small

125

computer monitor. The slide showed what had happened to the brain of a mouse that had spent weeks scampering on its little wheel, as many as twenty thousand rotations a day.

As the researchers watched the screen, tiny green dots appeared in the microscopic view of the mouse's brain. The dots were new brain cells tagged with a dye that made them glow bright green. There were hardly any green dots in the brains of the mice that had not exercised, but in the mouse that faithfully and voluntarily ran on its wheel, there they were, as clear as day—small green dots in the middle of his dentate gyrus. Exercise had prompted the birth of new neurons—neurogenesis. And the scientists, seasoned veterans all, could hardly believe their eyes, in particular Scott Small, in whose lab the new baby neurons were born. "To see those green dots light up in the mice," said Small when I spoke with him about that day. "To see it so clearly, new brain cells that came with exercise, it was impossible to ignore. My colleagues started putting on their sneakers."

Over the past several years, neuroscience has been on a serious hunt to figure out how to nudge our brains in the right direction as they age. Are there real things that can help real brains in the real world? Education seems to buffer the brain. And there are a lot of other ideas—and loads of exaggerated claims—out there. Sudoku? Deep wave meditation?

At this point, the most promising answer is exercise. In one rigorous study after another, exercise has emerged as the closest thing we have to a magic wand for the brain, the best builder of branches, baby neurons, and, along with education, perhaps, the mental padding of cognitive reserve.

Scientists have suspected for decades that exercise, in particular aerobic exercise, is good for the brain, just as it's good for the heart. Like all our cells, brain cells need oxygen, and the more our blood can spread oxygen around, the better. Blood flow is blood flow. What we're told to do for our hearts—keep our cholesterol and blood pressure under control to make sure our arteries are as nimble as possible, for instance—turns out to be just as good, perhaps even better, for our brains. And Scott Small, with his green dots, is at the forefront of what is now an all-out effort to figure out how exactly this works.

Energetic and talkative, Small, at forty-six, is doing what he can to maintain his own middle-aged brain. He loves a fast game of tennis and recently took up snowboarding. The question is, as he reaches his sixties, seventies, and even eighties, will all that moving around leave his frontal lobes as finely tuned as his forehand?

In the spring of 2007, Small published an extraordinary study that suggests that the answer is yes. The study, starting with animals, first divided forty-six mice into two groups. For two weeks, one set of mice was kept in cages with running wheels and the other without, after which time the researchers scanned the mice to see what was happening to the blood flow in their brains. In fact, Scott Small's lab was one of the first to develop the techniques to scan the brains of tiny mice, a step that not only lets researchers see what is happening brain cell by brain cell but can validate findings from sometimes difficult-to-interpret human brain-scanning studies. In this study, for instance, the mice were injected with a substance—now banned in humans—that clings to new cells and allows scientists to see precisely where new brain cells form. Then the researchers looked at microscopic slices of the mice brains.

Small and his colleagues found what they expected. The mice that had run on their wheels had increased blood flow in their dentate gyruses, those tiny sections of the memory-crucial hippocampus. The increase was there long after the mice stopped exercising, too, which meant it did not come from the transient boost in metabolism that regularly occurs while exercising. And right in the middle of those mouse brain dentate gyruses, Small and his team also saw those green dots of new brain cells. There were nearly *twice* as many green-dotted new brain cells in the exercising mice as in the nonexercisers.

For Small and his colleagues, it was a powerful finding, a clear sign that exercise was not only a potent producer of new neurons, as some earlier results had suggested, but also seemed to "selectively target" the brain's dentate gyrus—right in the middle of the brain's memory machinery—an area that appears to decline with the normal aging processes. This means that exercise may, in fact, help boost our memories as we age.

"The hippocampus is really a circuit of different regions," Small says, "and exercise targets this specific area of that hippocampus, the dentate gyrus."

While it's true that Small's green-dot study was tiny, it makes sense in part because it builds on solid research by other top neuroscientists. In fact, some of the first clear evidence that exercise boosts our brains came from the coauthor on the Small exercise study, Fred Gage.

One of the most well-known neuroscientists working today, Gage was the first to find that running—and running alone—could give birth to new brain cells. In the late 1990s, Gage and his colleagues, including Henriette van Praag, at the Salk Institute in La Jolla, California, decided to see what would happen if mice were allowed to run as much as they wanted, which usually meant four or five hours a night, or up to five kilometers.

Gage then put the mice to a classic test. He placed them in a tank of murky water and let them find a tiny platform hidden below the surface that they could land on. This is called the Morris water maze and is one of best ways to determine how smart a mouse is—a kind of mouse IQ test, if you will. Mice don't really like to swim, so when they're dunked they try as hard as possible to find the little platform. Those that locate it faster on subsequent dunkings are considered cognitively ahead of their peers.

What Gage and his colleagues found was that the mice that exercised the most were not only much better at finding the platform on the second and third tries but also had twice as many new neurons in their brains.

And where were the new neurons? Just where he and Small later found them in the Columbia University mice—in the middle of the dentate gyrus. "Our results indicate that physical activity can regulate hippocampal neurogenesis, synaptic plasticity and learning," Gage concluded in his 1999 paper. In later studies, he found that exercise woke up the newborn neuron machinery in elderly mice, too.

Of course, these experiments were conducted only on mice. But Gage, once called the "impresario of neuroscience," was determined. He was on the trail to find the roots and promise of neurogenesis.

100 Wrongheaded Years

Like most new concepts, the idea that an adult brain—animal or human—could actually grow new brain cells got off to a bad start, a prime example of how science moves ahead in fits and starts at best. Until recently, most neuroscientists had not budged from conclusions drawn in 1913 by Spanish researcher and Nobel Prize winner Santiago Ramón y Cajal, who confidently wrote: "In the adult brain nervous pathways are fixed and immutable. Everything may die; nothing may be regenerated."

It was an idea that seemed to make sense. But it was wrong. Gage himself has written about why it was so hard to believe that this idea could be incorrect in a 2003 article in *Scientific American*:

For most of its 100 years history, neuroscience has embraced a central dogma: a mature adult's brain remains a stable, unchanging, computer-like machine with fixed memory and processing power. You can lose brain cells, the story has gone, but you certainly cannot gain new ones.

How could it be otherwise? If the brain were capable of structural change, how could we remember anything? For that matter, how could we maintain a constant self-identity? Although the skin, liver, heart, kidneys, lungs and blood can all regenerate new cells to replace damaged ones, at least to a limited extent, until recently scientists thought that such regenerative capacity did not extend to the central nervous system, which consists of the brain and the spinal cord. Accordingly, neurologists had only one counsel for patients: "Try not to damage your brain because there is no way to fix it."

As far as anyone recalls, the first hint that this might not be so came from a young scientist at the Massachusetts Institute of Technology, Joseph Altman. As Sharon Begley writes in *Train Your Mind, Change Your Brain,* which traces the history of neurogenesis (and, interestingly, relates the latest neuroscience to the teachings of Buddhism), Altman was itching to test out a new technique that allowed researchers to tag newly formed DNA in cells with a radioactive substance.

In the early 1960s, he decided to use it to see if he could find any new neurons in the brains of adult rats. And he did. He then went on to find newly formed brain cells in the brains of cats and even guinea pigs. He published his findings in a scientific journal, but it attracted little attention and he soon transferred to Purdue University and dropped the still-too-controversial concept of neurogenesis.

The idea, however, did not go away. Studies in the early 1980s in songbirds, in particular canaries, found that they, too, created new neurons, even as adult canaries. Each spring, as the canaries learn a new mating song, new sets of neurons are created and migrate into their song-making brain area, which becomes correspondingly huge.

Then, in the late 1990s, Fred Gage found that the same thing happened in rats. Adult rats that lived in stimulating environments—with other rats, toys, and wheels—as well as rats that exercised, created many more new brain cells. He also found that exercise alone produced new neurons. Later, other researchers found new neurons in adult monkeys as well.

Next came humans, and for this study Gage teamed up with Swedish neuroscientist Peter Eriksson, who had obtained brain slices from older Swedish cancer patients who had been injected with a substance that would tag dividing cells. They were able to show that even adult humans were producing new neurons. And where were those baby neurons showing up? "We demonstrate that new neurons . . . are generated from dividing progenitor cells in the dentate gyrus of adult humans," Gage wrote when his research was published in the journal *Nature Medicine* in 1998. "Our results further indicated that the human hippocampus retains its ability to generate neurons throughout life." It was a study that changed brain-research forever.

Exercise Equates with New Brain Cells

And it did not stop there. Gage, along with Small, went on to extend the green-dot mouse study at Columbia to include humans as well. That was where Kevin Bukowski came in. Piggybacking on an exercise study that was being conducted by his colleague Richard Sloan, a behavior psychologist at Columbia University, Small decided to take a peek at the dentate gyruses of eleven people who were in

Sloan's experiment. (The Sloan study asked whether high-intensity exercise could cut down on markers of inflammation, which can harm cells. It did.)

After Small scanned the brains of the humans, he found pretty much what he'd found in mice. The humans, like Bukowski, who had exercised the most had twice the blood flow as the nonexercisers, and the increase occurred in that crucial memory area, the dentate gyrus.

What's more, the dentate gyrus blood flow jumped the most in those who became the most fit, as measured by their level of VO2 max, or the maximum amount of oxygen they took in as they exercised, the gold standard for measuring fitness. And that same most-fit group also improved the most in cognitive tests.

"We were not sure what we would see, but it was one of those days when the muses of science were smiling on us," says Small.

Because the researchers could not cut open Kevin Bukowski's head, and radioactive substances that tag new cells are now off limits for use with humans, the study did not technically prove that new neurons were born in Bukowski's brain. They can point to no green dots. Still, because of the striking increase in the level of blood flow—a measure that correlated directly with the growth of brain cells in mice—Small and others feel safe in saying that exercising appears to promote the birth of new brain cells.

"We can't validate the finding in humans, but by inference we can say that exercise drives neurogenesis," Small says.

That leaves, of course, the question of what difference any of this means to us. What's so special about a few more neurons? Can a few brain cells here and there hold off the assault of aging? Put another way, are a handful of baby neurons in something so small and obscure as the dentate gyrus really a good enough reason to turn off the TV and get out on the track?

In fact, when I went to see Gage at his California lab, this was the main question on my mind.

Gage's office is on the bottom floor of a row of concrete buildings that make up the Salk Institute for Biological Studies in La Jolla. The buildings, with their unadorned style, famously designed by Louis Kahn, make the most of their astonishing setting—on a high isolated bluff overlooking the Pacific Ocean. Inside are dozens of working labs that produce some of the most important biological research in the world.

Despite its prominence, the institute is a surprisingly informal place, with bikes leaning against walls and open-air hallways. After I finally found his office in the maze of concrete, Gage was relaxed and informal. Dressed in a short-sleeved yellow shirt, he was, at age fifty-seven, still trim and athletic and bounded up to greet me with a big handshake and a genial grin.

After we settled into his small office, I asked him, "Why *should* we care about these baby neurons?"

The question made Gage laugh.

After all, he has spent the last ten years working to prove that new neurons exist at all, an idea, he says, that has only recently reached a point of "growing acceptance." It takes a strong and energetic mind to take on the next challenging questions of what these new little neurons do, how they do it, and, why, in fact, we should care.

Gage, of course, has just that kind of strong and energetic mind and these are precisely the questions he is now addressing.

"The new brain cells are integrated into the existing circuit, no question," Gage told me. "But the question now is, how do they do it and why?"

At this point, he still shakes his head over how long it took to convince the scientific community that neurons are, in fact, continually born in grown-up brains. Doubts persisted, he said, because "for the longest time we thought that the brain was like a computer and if you threw a new wire into that existing circuit you would just screw it all up. Now we know that is not the case," he told me as we sat in his tiny office. "The brain is an organ. It is tissue that is changing all the time and it is regulated by our environment. Our brains are affected by what we do."

We now know, too, that the new brain cells—which are stem cells, the very earliest and most versatile version of cells—are primarily produced in that tiny area of the hippocampus, the dentate gyrus. We know that about half of the new cells die off and half survive. And we know that they are produced in a variety of ways. We get new neurons when we focus on a task that's highly complex or even when we're focused on a specific goal (such concentrated brain activity produces theta waves, the same kind of waves that are produced with meditation—so the claim that theta waves help our brains may not, in fact, be just hype).

And we know that exercise—regular exercise, which includes just about anything that increases heart rate and blood flow—leads to a boomlet of such babies.

"Just look at this," said Gage, wheeling his chair around to click on his office computer, where a slide appeared. One squiggle of magenta on the screen was the enlarged picture of a mouse hippocampus. On top of that was a sliver of dark blue, the dentate gyrus. Extending out of that sliver were dozens of branches—mature neurons. And scattered among those branches were tiny bright-green dots. The same dots—the same baby neurons—that made believers—and joggers—out of the workers in Scott Small's lab.

These new brain cells that I was looking at on the computer screen, Gage explained, had been produced in only an hour and a half in the brain of a mouse that had exercised. And seeing those green dots for the first time, I must admit, was impressive, inspirational, even. And this was just one small brain slice from one short moment in the life of one small mouse.

"The thing we have to remember is that neurogenesis is not an event, it's a process," Gage said. "And there's no question, physical activity makes new brain cells proliferate."

Details, of course, are still being worked out, but Gage is convinced—by his own work and that of others—that exercise produces new brain cells in a fairly straightforward way. When muscles contract, they produce growth factors, with names like VEGF and IGF. Normally, those growth-factor molecules are too large to make it through the blood-brain barrier, but for reasons that are still unknown, exercise makes that barrier more porous, allowing those growth factors, once referred to as Miracle-Gro for the brain, to get through and help stimulate the neurons. (The same thing has been shown to happen with serotonin, which is increased in the brain with exercise and also makes new brain cells grow.)

After that, things get fuzzier. The number of new neurons we produce, as Gage says, "has tremendous genetic variability." No one knows exactly how many we churn out overall. In all likelihood, it's a relatively small number, perhaps, Gage says, in the "single digit percentages," compared with the total number of brain cells.

So what exactly *are* they doing?

Bringing New and Old Together

Through a complex system of math modeling, backed now with more animal data, Gage has recently developed a new and fascinating theory about newly generated brain cells. He believes that the cells are crucial for our entire lives, and are doing nothing less than "helping us make sense of the world." In particular, he says, they "help us adapt to the new," to fold new experiences into our existing view of the world.

"If we were spending our whole lives in this room, we would not need new brain cells," Gage told me, gesturing toward the walls of his cluttered office. "But the new brain cells help us integrate the new with the old. Without them, we would never want anything to change because anything new would be too complicated."

When sensory input first comes into the brain—to put all this in its simplest form—it goes to its outermost layer, the cortex. The input then travels to the hippocampus, which consolidates information, memories, and learning. The hippocampus binds the varied sensory experiences together into a sensible chunk—and then sends that back to the cortex for long-term memory storage.

But before the information even gets to the central hippocampus, it is first filtered by the gatekeeping dentate gyrus, which is thought to perform an opposite task—it breaks sensations into even smaller pieces. It is, as Gage puts it, "a pattern separator." Brain cells in the dentate take note of subtle differences and similarities—a leaf a bit greener, tea slightly hotter. Mature brain cells in the dentate encode those minute differences and pass them on to the hippocampus.

So how do baby neurons fit into all that? Initially, Gage believed that the new cells—since they're formed in the dentate—must somehow help it do its job, that is, break up information. But, Gage told me almost proudly, "I was really wrong."

Instead, it now appears that new neurons may actually act to tie disparate information together—and place that information in a specific time frame. Gage now believes that new neurons help us make associations. If we hear a Beach Boys song and smell the salt from the beach, those two impressions—Beach Boys song and salt smell—will be forever tied together in time and place. In fact, the more neurogenesis you have, Gage says, "the more you link together things that are different" into a pattern that will hang together in your brain.

Our memories are notoriously unreliable, in part because we are constantly pulling up old memories and "retagging" them with new information, then restoring the memory in a modified form. Baby neurons, Gage believes, help us with that process, tying together different sensations that occur at the same time—and helping us fit the new with the old, the song we know with the sand we are sitting on.

With chronic stress, new neuron production is slowed or grinds to a halt. To explain this, Gage uses an example of a soldier in Iraq with post-traumatic stress disorder, PTSD. Imagine, Gage said, that "you have a soldier in Iraq and he is under chronic stress and therefore not producing new neurons. [Then there's a] stark event, say, he sees his buddy's head blown off."

If neurogenesis were occurring, even that stark event would—when the soldier recalled it later—be *retagged* with new information and the soldier might be able to soften the memory by mixing it with more random—and gentler—everyday sensations and information before it is restored as a less upsetting memory.

"Neurogenesis links different things together and that helps us generalize experiences and rationalize them," Gage explained. Without a stream of new neurons, Gage believes, such a memory would be stored only in mature brain cells and there it would stay—"the event would stay as the event," as stark and real as it was.

Gage thinks this is one way that talk therapy might work. If we recall bad memories in a safer environment—and if we are not under stress and new neurons are being produced—those memories will be mingled with gentler thoughts—nice office, calm therapist, flowers on the table—and that may be what helps us make sense of—and live with—some of our most disturbing memories over time.

Gage has also developed a model of how all this might happen. In essence, when input comes into *mature* neurons, it's encoded. But that encoding is then quickly halted by a neurotransmitter that inhibits brain activity, GABA. If the encoding were not stopped at some point, the older neuron would be constantly readjusting to new information.

But new baby neurons are set up quite differently. For the first seven days of their life, before they have formed connections with other neurons, Gage says, they are actually *excited* by GABA, rather than shut down. That means that as they are born, they will soak up some GABA from nearby mature neurons, and get excited at the very nanosecond that older neurons are both being activated and then shutting down.

As a result, the new baby neurons encode information from all their neighboring mature neurons—salt, sand, song—tying it all together in time in a mixed memory that stays with us until it resurfaces and is remixed and restored again. The new neurons have time-stamped memories.

This idea is still unproven, of course, but it is—perhaps not surprisingly, given its source—an elegant and compelling one.

Gage believes this may be the way that neurogenesis can alleviate depression, helping us maintain interest in our world. When we get sick we often become immobile, and with "that lethargy we stop producing new neurons, leaving us both less cognitively aware and depressed," he says.

After all, he adds, "what is depression but a lack of interest in the new, the feeling 'Is that all there is?' We need new neurons to help us adapt to the new, to put it in context. Sometimes to get excited about things, you have to recognize how this cool new thing is like other cool things we knew about in the past. Neurogenesis helps us do that."

In Gage's view, in fact, the whole system might have developed to *allow* us to deal with the new. While this, he admits, is getting into the often fuzzy area of evolutionary theory, again, it is such an interesting thought that it seems worth mentioning.

"Just think about it," Gage said. "As soon as the primitives walked out toward the savanna, the walking would have stimulated the production of new neurons that they would need to prepare for their new environment, to adapt to it and integrate it with their old environment."

There is little question that in general neurogenesis declines with age, sometimes starting in middle age. But we also now know that, as Gage puts it, "the cells are there and we can reactivate" the process.

But we need to get up out of our chairs. Indeed, Gage is such a believer in the power of exercise to keep those baby brain cells blooming that he runs "a lot" so he will then be able to play squash "with the young guys" four or five times a week. He and his wife try to walk whenever they can. And he recommends we all at least try to do *something* for thirty minutes a day—to get that dentate gyrus up and pumping and get our dose of fresh new neurons.

"This is not about finding a drug," he said. "This is a lifestyle thing. The drug companies don't like to hear that, but we can affect what happens in our brains with what we do."

Boosting Brain Volume

There's also emerging evidence that exercise helps the brain in more global ways as well. It's still uncertain how much exercise a brain needs, or, as one scientist said to me: "In exercise, we don't yet know the dosage." By and large it appears that anything that increases your heart rate helps.

But that doesn't mean you have to sign up for the New York City Marathon. And for that bit of good news, we can thank Art Kramer, the neuroscientist at the University of Illinois at Urbana-Champaign.

Kramer is interested in the exercise-brain connection not only as a top scientist but also as a middle-aged man with a worrisome family history. Always a bit of a jock, Kramer boxed as a young man, then moved on to running and track, and now tries to get onto the stationary bike when he can. He also plays a game of squash that smashes twenty-year-olds.

Still, his father died young, and if he did not take drugs to control it, his cholesterol would be about 400. Like most of us reaching the middle of our lives, Kramer is concerned. Is he doing enough? Should he exercise more? Does any of this make any difference?

"It doesn't matter how long we can live, it matters how long we can keep going functionally," Kramer pointed out, quite logically, when I spoke with him recently. And so what are we to do to keep functioning well, and prove, with solid science, that what we're doing really works? So far, Kramer, doing his part, has found encouraging news about fairly moderate levels of exercise.

In one of his latest studies, published in 2006, for instance, Kramer and his colleagues found that those over age sixty who did regular stints of aerobic exercise for six months had increased brain volumes in their frontal lobes' gray matter, which includes the neurons, and in the white matter of their corpus callosum, the nerve bridge that connects right brain to left brain—and whose age-linked deterioration has been associated with slower thinking.

The exercise in this case was a fairly mundane program of brisk walking. Those who spent about an hour walking around a gym three times a week—at a pace of three miles an hour—had brain volumes of people three years younger.

"Significant increases in brain volume, in both gray and white matter regions, were found as a function of fitness training for the older adults who participated in the aerobic fitness training but not for the older adults who participated in the stretching and toning [nonaerobic] control group," Kramer concluded in the study in the *Journal of Gerontology.*

"These results suggest that cardiovascular fitness is associated with the sparing of brain tissue in aging humans. Furthermore, these results suggest a strong biological basis for the role of aerobic fitness in maintaining and enhancing central nervous system health and cognitive function in older adults."

That impressive study came after a stream of similar research by Kramer and his colleagues. One study in 2003 found that those over age sixty who exercised regularly—again, that meant aerobic exercise such as running or walking quickly—had less brain tissue loss than non-exercisers. And in a study published in 1999 in the prominent science journal *Nature,* Kramer reported that a group of 124 relatively unfit people over age sixty, after walking rapidly (17.7 minutes per mile) for forty-five minutes three days a week (and especially those who managed to get up to a mile-long loop around the university at a good clip of 16 minutes per mile) were much better at complex tests, in particular those that involved "task switching," the same frontal-lobe challenge that faced Mark Moss's middle-aged monkeys. For the humans in Kramer's study, this test also involved rapidly answering questions such as "Is this an odd or even number?"

The exercisers were also better at focusing and ignoring irrelevant information. Such frontal-lobe executive functions, as we've said, are crucial for a whole range of everyday activities, especially when we have to do two things at once.

Kramer's studies also mirror solid tests in animals over recent years, including one at Oregon Health & Science University that found that monkeys that ran on a treadmill for five days a week for twenty weeks had much higher blood volume in their brains' capillaries than sedentary monkeys—and it was the oldest and least fit monkeys that had the biggest gains. As Kramer says: "We know from all this research that there are a few good things for the brain— and one is exercise."

Kramer says that he, too, is still not completely sure how all this happens in the brain and is now probing deeper to see if he can find more clues. It could be the dentate gyrus and it could be a combination of effects in the brain. "What all this precisely means on a molecular level we just don't know but we can speculate," Kramer says.

"We want to know what the nature of the volume change is," he told me. "Is it the growth of blood vessels or the number of synapses or white matter or gray matter? We just don't know yet."

When I last spoke with him, he had decided to extend his human walkers' study and then take their blood to see what genes are modulated and "who does better and why." He will also check the participants' blood for the presence of markers for inflammation linked to cardiovascular problems and possibly Alzheimer's (a number of conditions, including obesity and smoking, are now thought to produce a kind of low level of chronic inflammation in the body, which over time may wear down cell defenses and lead to disease).

If he could, Kramer would love to do a spinal tap on volunteers and look for nerve growth factor, the Miracle-Gro, in the nervous system of study participants, but that's not something one does with living humans. "We can't get a slice of their hippocampuses, either," he said, a bit sadly. Still, the thought that something as simple as exercise can have real benefits for our brains is just the kind of optimistic thought that may very well appeal to our positive-seeking middle-aged brains.

And it's an idea that now makes perfect sense to people like Kevin Bukowski. Having ignored his brain for years—much like the rest of us—he is now giving it much more respect. Like most of us in middle age, Bukowski is busy. He has a demanding job, assisting and coordinating scientific trials at a major medical center. He has a seven-year-old daughter. He is taking care of his mother. But at age forty-seven, he feels somehow sharper and "calmer . . . there is a maturity factor there now.

"I just feel now at my age that I am doing a lot. But I feel I can really handle it all now and that makes me feel good. It is kind of surprising but here I am at middle age and it's not bad; in fact, I feel more secure knowing I can deal with all this."

But he is also now convinced that to maintain all that, he has to stay on the treadmill. And so he is setting out to do what he can. He is training his middle-aged body and brain—for a triathlon.

10 Food for Thought

And a Few Other Substances, as Well

In 2008, the giant British drug company GlaxoSmithKline announced it would pay $720 million to a small Massachusetts biotechnology company that's been working for years to prove that we can extend our lives by drinking a lot of red wine.

Specifically, the company had been trying to develop a pill with a high concentration of resveratrol, the ingredient in red wine that some believe may hold a key to keeping our cells, including our brain cells, healthy and nimble.

But Glaxo was not paying for something that's known to work in actual humans. It paid $720 million to a company whose work so far had shown that resveratrol can extend the life of yeast and, to a certain extent at extremely high doses, rodents.

Is this madness?

Clearly, we want very much for it *not* to be madness. The idea that something as simple as the wine we drink—or the food we eat—can make us think better and live longer is highly seductive. Indeed, it's hard to think of a topic more steeped in hope—and hype—than the brain and food.

As I write this, I've just returned from a Whole Foods supermarket, where a young woman—hired as a kind of modern-day circus barker—was handing out free samples of a new bottled tea. "Try it! You'll love it!" she nearly shouted to shoppers as they pushed their carts. "It's got MORE antioxidants than green tea! *More antioxidants for your brain!*"

I took one of her tiny plastic cups of tea and in one gulp drank it down. Was I smarter? Sharper? Was my middle-aged brain better braced with those antioxidants sloshing around?

Hmmm . . . well . . . maybe.

Uncertain Connections

Even those who study all this for a living are confused. While a consensus has emerged over the benefits of exercise and, to a certain degree, education, agreement on what works beyond that falls off a steep cliff.

"The message seems to change every year," says Mark Moss, the neuroscientist at Boston University, with a sigh.

Part of the problem in talking about food and the brain is that, on one level, it's a no-brainer. Who hasn't heard a parent tell—or been a parent who's told—a child, "Eat your fruits and vegetables!" or "Fish is brain food!"? Rules about which foods we're supposed to eat or not eat have been with us pretty much from the beginning. Even the sacred Hindu text the Bhagavad Gita, in which a young warrior is advised by an all-knowing god Krishna, talks about what to eat.

"There are three kinds of food as well," Krishna advises. "Here are the distinctions among them:

"Foods that are the *sattvic* are drawn to promote vitality, health, pleasure, strength, and long life, and are fresh, firm, succulent, and tasty.

"Foods that please the *rejasic* are bitter or salty or sour, hot or harsh or pungent, and cause pain, disease, and discomfort.

"Foods of the *tamasic* are stale, overcooked, tasteless, contaminated, impure, filthy, putrid, and rotten."

So there you go, diet advice from 5000 B.C.

With all that, we certainly think we know what we're supposed to eat. We know too much refined sugar or grain or salt is bad for our general health, as are the wrong kinds of fat.

We may ignore all that, but we do know it. What few people realize is that most nutritional advice is based on one particular kind of research—large population studies. These wide-swath studies have, in fact, found that diets high in fiber generally protect against colon cancer and those high in vitamin C reduce the risk of stroke. Researchers can trace the amount of fat in the diet to the rate of breast cancer in a population and watch those rates change as people move from one food culture to another. Japanese women do not suffer from the same rates of breast cancer as American women do, but after just two generations of living here, Japanese-American women, on average, have the same breast cancer rates as any other ethnic group. Similar population—or epidemiological—studies have repeatedly found that high blood pressure is associated with an increased risk of heart disease and that obesity is linked to many chronic diseases.

Still, if you peer closer, such correlations can disappear. Studies that keep track of specific foods and specific people often fail to find any effect at all. Just a few years ago, the now-famous Women's Health Initiative, the kind of giant, randomized, controlled clinical trial that's considered the Rolls-Royce of research, failed to confirm that a low-fat diet prevented breast cancer, for instance.

Long-term studies of the antioxidants vitamins E and C and beta-carotene that looked at their effects on individual people also found they do not prevent heart disease, and similar studies failed to confirm that diets high in fiber ward off colon cancer.

Certainly, a few foods have been directly linked to health. If we have a serious deficiency in our diets, adding a specific nutrient helps—folic acid for pregnant women to prevent neural tube defects in babies, for instance.

But our modern Western diets, as junk-food-filled as they may be, are also astonishingly varied and rich, with low-cost, fortified, and abundant food. Most of us get what we need and do not have serious deficiencies that can be easily corrected by adding this or that nutrient.

In light of all this, can we really expect that a glass of wine or a plate of spinach will make any real difference? Or is our fate prewritten in our genes and we're just fiddling around the edges or, worse, fooling ourselves?

Blood-Brain Barrier

And given such a dismal track record of food and overall health, can we possibly know anything about food and the brain? Can we figure out what we should put in our mouths to nudge our complicated neurons?

Until recently, this question was not even asked. For years, scientists believed diet had little impact on our brains because they thought most nutrients didn't cross the blood-brain barrier. The blood-brain barrier is real. Cells lining blood vessels in the brain are packed close together to keep out certain large molecules and maintain a chemical balance. Some bacteria are barred, for instance, and infections in the brain are rare for that reason. Paula Bickford, a neuroscientist who studies nutrition at the University of South Florida and the James A. Haley Veterans' Hospital, says that for a long time most scientists believed that even vitamin E did not cross the barrier. In one study Bickford conducted, in which she gave high levels of oxygen to rats to see if it would induce damage, one reviewer of her work asked her why she was wasting her time because "nothing could affect the brain." And that, Bickford says, "was just ten years ago!"

And there were other roadblocks as well. Until recently, most believed the brain was on a downhill slalom course from our mid-twenties, losing as many as 40 percent of its cells as it aged. Why bother worrying about a brain that is programmed from the get-go to decay? Did anyone really think another forkful of carrots was going to stop that from happening?

Now, though, we know that much of that doesn't hold up. The brain does not lose large numbers of neurons as it ages. Nutrients do cross the blood-brain barrier and are, in fact, essential for the brain. As a result, there's now renewed interest in figuring out how we can tweak our blood—stir in some growth factors from exercise or maybe some special ingredient in this food or that—to benefit our brains.

As Mark Moss in Boston says, "We never thought what was happening in the body was getting to the brain. We thought that the brain was protected. But we are finding that the blood-brain barrier can be breached. Believe it or not, the circulation people never talked to the brain people and now we are talking . . . and it's a big, big deal."

By nature a conservative lot, scientists don't like to put their money on something until they understand *how* it might happen, the mechanism. Scientists are more willing to look at the food-brain link now because they've identified ways it could work.

And what are the mechanisms by which food might bolster our brains? They involve the same substances touted for years in terms of overall health—antioxidants such as vitamins C and E, along with anything that acts as an anti-inflammatory, from fish oil to aspirin.

Why should we go there again? It's a legitimate question. Some suggest that we need to pay even more attention to the brain and food because the brain uses so much energy (at rest it uses 10 percent of the body's oxygen and in mental activity up to 50 percent), leaving it both sensitive to and in greater need of certain nutrients. Others don't go that far, but say that we've oddly missed the boat by thinking that the brain does not react to what we eat.

"The brain is not *uniquely* sensitive but it is *sensitive* to what we eat," says Bickford. "And now we're realizing more and more that what you eat can affect cognition."

Antioxidants and Inflammation

The theory of antioxidants and aging has been around now for more than thirty years, and it's intriguing that it's now at center stage again—in research into the brain as it ages.

The idea goes like this: When our cells (including our brain cells) burn oxygen to make energy, a byproduct called a free radical is produced and thrown off. A free radical is essentially a molecule that's missing an electron. And because it lacks that electron, it is unstable and wants to steal electrons from other cells. It grabs those electrons where it can, causing damage to other cells willy-nilly in the process. That damage is called *oxidative* stress and many believe it's one of the main culprits in normal aging.

So eliminating as many free radicals as possible seems like a good idea. When we're young, free radicals are often neutralized by another molecule, an antioxidant, in a continual repair program. But as we age—for reasons that are still not fully understood—that process becomes less efficient; the antioxidants can't keep up with the hordes of free radicals (which are also produced by exposure to environmental insults such as pollution, ultraviolet light, and radiation), and more brain cells are left dented and nicked.

Then, a few years ago, another potential aging mechanism was identified: inflammation. Inflammation occurs when the body is injured and white blood cells rush in to do repairs. With that incoming surge, there can be collateral damage. Like fire engines driving up on a nearby lawn as they speed to a house to put out a fire, the cells rushing in to help out often harm surrounding healthy cells. And if the brain or the body is in a state of chronic low-level inflammation, there's likely to be a buildup of damage. Unable to cope, cells—including brain cells—begin to shut down and die. Long-term inflammation (which many now believe can also come from chemical processes that accompany obesity or even excessive stress) may contribute to a number of chronic diseases, from heart disease to Alzheimer's.

So, if the evil twins of brain aging are oxidative stress and inflammation, the questions are: Will eating foods high in *anti*oxidants or *anti*-inflammatory agents make a difference? Does it have to be real food or can it be a supplement or a vitamin pill? And can we make an impact on our very complicated brains if we start gorging on antioxidants and anti-inflammatories at middle age, or is that too late?

No one thinks a lack of antioxidants or too many inflammatory foods are the root causes of neurodegenerative diseases. But nearly everyone now thinks they are, at the very least, somehow part of the normal aging process in the brain, and therefore possible targets for intervention.

"My own view is that antioxidant damage and inflammation are in the background and they make cells more susceptible to other insults such as neurodegenerative diseases," Bickford says. "If we get insults like that, we are less able to function at 100 percent and make repairs, then we start to see failures."

Normal aging in the brain, many believe, comes about in part because repair mechanisms slow down. That process is complex but does seem to include antioxidant damage and inflammation, and both, Bickford says, "can be affected by nutrition," including what we eat in middle age.

"When a cell is dead it's hard to bring it back to life," she says, "but certainly in middle age and up until the point of no return, I think there is room for repair."

Brain Foods

With all this new information, science has now embarked on a merry chase to figure out which foods give our brains the biggest bang for our buck. Several years ago, the U.S. Department of Agriculture, with researchers at Tufts University in Boston, developed a method to screen and rank foods for their antioxidant capacity. The list is known as ORAC, for oxygen radical absorbance capacity, and it's one of the main reasons we suddenly started eating bowls full of blueberries. Just to give you the top twenty-five, the list goes like this:

Prunes, raisins, blueberries, blackberries, garlic, kale, cranberries, strawberries, raw spinach, raspberries, Brussels sprouts, plums, alfalfa sprouts, steamed spinach, broccoli, beets, avocados, oranges, red grapes, red peppers, cherries, kiwifruit, baked beans, pink grapefruit, and kidney beans. (A key to this is color—generally, darker is better.)

Down the list a ways are carrots, coming in at forty, and tomatoes, at forty-two.

So does that mean we have the answer? Have a hearty breakfast of prunes and kale and, if you're not up for a glass of red wine (red grapes) that early, a cup of black tea (also high in antioxidants) and call it a day?

Not entirely. Despite a growing belief in the potential of foods such as blueberries and spinach to help stave off aging in the cells of the brain as well as the body, it's important to remember that, in terms of the brain, there have been no long-term clinical trials in humans to test all this. Not one.

Most smaller trials of vitamin C or E or ginkgo biloba and cognition have been mixed at best. There are ongoing large human trials to see if curcumin, which is the turmeric found in a lot of Indian food (Indians have a lower rate of Alzheimer's than some other populations), or even caffeine (which we all know can give a needed jolt to nerve cells), can help ward off Alzheimer's. But there are no full-scale controlled trials to figure out whether what we eat can help us remember what movie we saw last night.

"There's always a push to address a disease, but with normal aging we always have the question, what is the product that a drug company can sell?" says Bickford. Consequently, there is less funding for research into normal brain aging, which, she says, is a shame.

Still, that does not mean that science has given up. After a recent summit meeting of cognitive-aging scientists organized by the National Institute on Aging, I had dinner with two of the leading neuroscientists working on aging and the brain, Denise Park and Laura Carstensen. As I joined them at their table in at a Washington, D.C., restaurant, they were talking about the subject that has now taken center stage—possible interventions to improve our brains. Many, including those who study rats, monkeys, genes, and humans, say they now believe more strongly than ever that eventually it will be possible to slow down or even halt or reverse the aging process in the brain.

Most scientists at the conference spoke glowingly of exercise, the current star in brain-aging research. But others talked about the anti-aging potential of a whole range of substances, most of which are antioxidants or work as anti-inflammatories. Some touted spirulina, a kind of algae, which is what fish eat to get all those good omega-3 fatty acids, which can reduce inflammation. Why not skip a step and just eat the algae ourselves?

Others spoke confidently of "druggable" targets in the brain, where it might very well be possible to boost repair processes by taking a pill—nutrient-derived or otherwise. A growing number of neuroscientists have gone so far as to form their own companies to both research potential agents and, presumably, cash in when they find something that works.

At this point, each neuroscientist seems to have his or her own pet formula. Some continue to believe that, for females at least, estrogen might work. For years, test tube and animal studies have consistently found that estrogen seems to nourish the brain, making connections grow in areas like the hippocampus, where memories are formed.

But that does not mean that adding a dollop of estrogen to an aging human brain is necessarily a good idea. Indeed, any recommendation of estrogen these days is messy at best. First, the big studies appeared to find that hormone therapy (estrogen) decreased dementia, then later, more rigorous studies found that estrogen not only didn't help keep us mentally sharp but actually *increased* the rate of dementia, not to mention raising the risk of breast cancer and strokes. Some now argue that even those results were flawed because they did not study women who were young enough—at the start of menopause, for instance—when their brain cells were still stable and healthy enough to soak up estrogen's bounty. But no one has yet been able to prove conclusively that there is such a critical window. And given its known risks, estrogen is now shrouded in confusion and fear.

Some scientists, such as Roberta Diaz Brinton, a neuropharmacologist at the University of Southern California, are trying to develop plant-based estrogens that target only the brain and not the breast. Having spent years watching estrogen's effects in the lab, Brinton, along with a fair number of others, believes the hormone is essential to stabilizing and maintaining a neuron's energy metabolism and "keeping the cell in a survival mode." Some small human scanning studies have found increased metabolism in the frontal lobes of women given estrogen.

When I spoke with Brinton, I told her of a friend who was convinced that her brain at menopause had gone haywire, but then, after a few years, it came back and seemed to work fine. In Brinton's eyes, this is entirely possible, because menopause is "all about getting your brain to adjust to a lower level of estrogen," and once it does, things can improve.

In middle age, "The female brain changes from a reproductive brain to a nonreproductive brain," Brinton added. "Just as all those neurochemical circuits came into place at puberty . . . later, at perimenopause, the brain is upregulated for estrogen and then downregulated . . . and circuits, the bridges, are being dismantled. Some are left behind, but the system, the pathways, are being dismantled. After all that the brain is left unreceptive to estrogen. Menopause is all about shutting the brain off to estrogen and it's an adaptive process."

Still, the underlying message about estrogen, menopause, and the brain is—again—variability. "There's a spectrum of responses that probably depends on how estrogen-dependent a woman is. As estrogen levels plummet back to prepubescent levels, some women get confused and unfocused and are generally miserable until brain chemicals stabilize. Others say, 'Hey, What's the big deal?' "

In the end, though, researchers like Brinton are not giving up on estrogen, despite its current bad reputation. Encouraged by recent reports that the new kinds of antidepressants such as Prozac work in part by increasing neurogenesis (the birth of new brain cells), some other scientists have formed research companies to look into those drugs, seeking the right combination.

Will Animals Point the Way?

It used to be that serious science lived in its own cloistered world. True science was undertaken for the sake of knowledge, and any legitimate researcher who openly talked of devoting precious lab time to making something for real people in the real world was not taken seriously. Certainly, anyone who even suggested they might want to *sell* what they were studying was quickly relegated to the snake-oil club. But not anymore. "There's been a real shift," Park said as we ate dinner that night in Washington. "Everyone is openly talking about interventions, even drugs."

One major reason for that shift, of course, is that scientists now believe that the brain does *not* undergo complete disintegration as we age. It's now known that we do not, under normal circumstances, lose large numbers of neurons. And if our neurons remain mostly intact, that means we may not need a complete renovation. Perhaps we can just do the kitchen, or a downstairs bathroom. If we don't have to actually bring dead neurons back to life, but instead simply jump-start a nerve signal here or there, maybe we can figure out how to do that. Maybe it *is* possible to keep the brain running much longer at its tough-minded, middle-aged level by just tinkering a bit around the edges.

And maybe some of that tinkering could be accomplished simply with the foods we eat. "This is not the way of big pharmacology companies," says Bickford, who was trained in pharmacology. "But maybe because I am a child of the sixties, I find the idea that there is a simple way to do this just such an optimistic thought."

That optimism, given the lack of human data so far, is being fed by an increasingly rich trove of research on the aging brains of animals. What's true in a rat is not necessarily true in a human, but some of the animal results are intriguing.

"We don't know everything but we do know a lot," says Bickford. "In many ways, just over the past few years, really, nutrition and neuroscience have come of age."

A number of basic test-tube experiments have shown that brain tissue taken from older animals is more sensitive to oxidative stressors than similar tissue from young animals. By middle age, there are already indicators of increased inflammation in the brains of animals. But food seems to help. In animal studies, Bickford found that older rats fed a diet of dried spinach learned new tasks much faster than those fed plain rat chow. Rats fed diets enriched with blueberries, spinach, or spirulina had less brain-cell loss and improved recovery of movement following a stroke.

James Joseph at Tufts University and his colleagues Mark A. Smith and Barbara Shukitt-Hale have done dozens of experiments trying to zero in on what it is exactly in a blueberry that might be helping the brain. In one study, older rats (about sixty years old in human terms) that were already showing cognitive decline were fed extracts of two of the fruits highest on the ORAC scale, blueberries and strawberries, and did better on cognitive and motor tests. And when their brain tissue was examined, it had lower levels of markers for oxidative stress and inflammation.

In another Tufts experiment, after four-month-old mice with a clinically induced form of Alzheimer's ate blueberry extract, they did as well in memory tests in middle age as mice that did not have Alzheimer's and considerably better than mice with dementia that did not eat blueberries. And that was true even though the brains of the Alzheimer mice that had eaten blueberries and those that had not had the same amount of plaque damage in their brains. What's more, the blueberry-eating mice also had increased activity in molecules that are part of the learning and memory pathways. The mice were some-how protected. Is this cognitive reserve in mice? Mouse escapees?

Other Tufts studies suggest that blueberries can even increase the birth of new neurons in the dentate gyruses of older mice, the same part of the brain that's involved in memory and the same section of the hippocampus that's affected by exercise in mice and humans. The studies have so convinced James Joseph, he has taken to calling blueberries "brainberries" and starts his day with a big cup of them himself.

One of the most famous studies involving animals and nutrition was done by Carl Cotman at the University of California at Irvine and his colleagues, who found that beagles fed a diet of fruits, veg-etables, and vitamins and allowed to exercise could, even in old age, learn new tricks faster than dogs that did not have such good habits. (The dogs who did the best had their diets fortified mostly with anti-oxidants, including tomatoes, carrot granules, citrus pulp, spinach flakes, and vitamins E and C.)

And this may be a good example of brain maintenance. Dr. William Milgram, the study's lead author, said that even a relatively dumb dog, whose name was Scamps, appeared to do better after two years of the fortified diet than other dogs as they aged. "What hap-pened," Milgram said, "was that he remained the same while the dogs in other groups showed the expected deterioration."

Monkeys and Myelin

At the moment, one of the most ambitious efforts to pin down what food or substance might work is under way in Boston, where Mark Moss is trying to figure out—in a rigorous scientific way—how to slow down aging in the brains of middle-aged monkeys.

Moss, as we've mentioned earlier, thinks some normal aging (from memory loss to balance problems) can be traced to the gradual disintegration of the white matter in the brain, the fatty coating on the neurons' long spiked tails that sends signals across the brain. If the white matter—or myelin—is damaged, signals slow down or even get lost.

It's clear that as white matter increases as we age—into our fifties and sixties—we seem to get smarter, to see a more integrated picture of the world. But at some point, the repair process of the myelin breaks down. No one knows when that tipping point is, but many, including Moss, believe that somewhere during middle age, we end up "on the cusp" between effective repair and the beginnings of decline. Could we jump in at that point and slow or halt the aging process or boost repairs with some kind of food or substance?

When I first spoke with Moss at his Boston office, he had assembled a team of biophysicists who were completing sophisticated mathematical modeling to determine which substance might work best. Research on monkeys, like that on humans, is both expensive and difficult, with complex rules to protect the animals. So before proceeding, Moss wanted the team to figure out which agent had the best shot of helping.

At that point, his list was long and included antioxidants such as grape seed extract, anti-inflammatory agents such as spirulina and aspirin, and even statins, which lower cholesterol levels and may help blood vessels in the brain as well.

When I last spoke with Moss he told me that they had finally decided to test a number of things. One group of monkeys will engage in rigorous, carefully measured exercise (including a "very large hamster wheel"), and have their blood pressure and heart rates monitored. Another group will be given a food-based antioxidant, probably grape seed extract. And a third group of monkeys will be given an anti-inflammatory or a statin. The trial will go on for three years (equivalent to about a decade in humans) and then the monkey brains will be scanned and examined to see how their white matter fared and whether, even if we start at middle age, we can make a real difference in how brains age.

"I think you will see a rapid application of all this," Moss told me. "I think we will go the doctor, who will say, 'Hey, Mr. Jones, your kidney function is fine but your white matter voxels are a little low here and there.' We already have the technology to do that and I think we will be able to intervene. I believe it will happen because baby boomers are not going to take no for an answer—we want to keep going and going and, by god, we will do it."

There are those who can't wait. They see the animal data and simply can't resist. Paula Bickford eats a balanced diet but also takes spirulina and fish oil as well as a concoction she's cooked up herself (which she believes may stimulate stem cells in the body to do better repair as we age) that includes green tea and, of course, blueberries. Even Mark Moss, who scoffs at much of this, told me, a bit sheepishly, that he had increased his run to five miles a day and was—why not?—"also taking some grape seed extract."

Indeed, the idea of using a range of substances to boost our brains is no longer a fringe thought. While some worry that brain enhancement in any form might only increase the divide between the haves and the have-nots because some will have access to such substances and some may not, it is an idea that is increasingly discussed out loud.

Late last year, none other than *The Economist*, the normally staid British news magazine, had a half-page editorial backing the use of certain drugs that can work as cognitive enhancers, saying that "such drugs promise to do a lot of good." If scientists, for instance, used such drugs to help them focus and "unravel the mysteries of the universe, so much the better."

"Some worry about the unfair advantage and peer pressure that comes from these drugs," the magazine went on. "[But] is it 'natural' to prop up the aging body with a nip and a tuck, but to restrict help for the aging mind to brain-training on the Nintendo? It may even be that like Viagra, society largely welcomes the arrival of a chemical that does, far better what omega-3s, ginseng, vitamins and all the other quackery have failed to do. Unless of course, you want to outlaw double espressos, too."

Indeed, even before *The Economist* weighed in, another British magazine also went to bat for brain boosting. The leading scientific journal *Nature* conducted a survey of its scientifically inclined readers and found that one in five of the fourteen hundred respondents were already taking prescription drugs such as Ritalin and Provigil to increase their concentration or focus. The drugs are approved for various disorders, such as attention deficits or narcolepsy, but can legally be used "off label" to boost concentration.

Low, Low Calorie Diet

Adding something to your system is not the answer for everyone, though. One of the most extreme of the current food-researcher guinea pigs is Mark Mattson, chief of the neurosciences lab at the National Institute on Aging. However, his idea is not taking *more* of something but much, much less.

For the past twenty years, Mattson has been researching the idea of severely restricting the amount of calories consumed, which is the only intervention that has consistently been shown to lengthen the life span of everything from worms and fruit flies to mice.

Although there's no good solid data yet on humans, Mattson is doing his part to help. Since graduate school, he has been cutting way back on food. He now eats only two thousand calories a day, which, he told me, is "low for an American male" but "not that low." Mostly, he eats complex carbohydrates and "lots of fruits and vegetables." The father of two (who eat like the rest of us), he also coaches the high school cross-country team and "runs with the kids." All this has left him stick-thin. At five feet nine inches, he weighs only 125 pounds. Indeed, when I saw him recently, he was wearing a striped shirt, and some of the stripes looked wider than he was.

A good-natured man of fifty, Mattson is aware that not everyone could do what he does, but for him, he says, "it works."

So I asked him, "What did you have for breakfast?"

"Well, actually, I didn't eat breakfast. Normally I don't," he answered, thereby swatting away a century of nutritional advice.

Mattson has no idea if his low-calorie diet will prolong his life. So far he's fine and has "no diseases." But a one-person study sample is hardly science. "Yes," Mattson added, laughing, "I am an example of one."

In fact, no one knows how long humans can live. Clearly there is some built-in genetic program for all animals; otherwise we would not have such natural variety. The average fruit fly lives for only 2 months, an elephant for 70 years, and a turkey buzzard for 118 years. Why is that? Is it the fast metabolism of the fruit fly compared to the elephant, with humans falling somewhere in between? No one knows. The longest-living human we know of in recent history was a French woman, Jeanne Louise Calment, who died in 1997 at age 122. (She was a smoker who loved chocolate, poured olive oil on all of her food, took up fencing at age 85, rode a bike until she was 100, and lived on her own until age 110. Of course, as a French woman, she spent a lifetime sipping red wine resveratrol, too.)

There is now a growing subculture of scientists such as Mattson and other dedicated ultrathin folks who are trying to see if they can extend their lives—and keep their brains whirring along at high levels—by eating less—a lot less.

The idea has been around since 1935, when scientists at Cornell University found, pretty much by accident, that rats that ate less not only lived longer but also had fewer chronic diseases as well. Since then, a steady stream of animal studies has repeatedly shown that caloric restriction, which generally means reducing normal caloric intake by about 30 percent, can extend the life of animals as much as 30 to 40 percent, as well as delay or prevent such chronic diseases as diabetes and atherosclerosis and neurodegenerative diseases such as Alzheimer's, Parkinson's, and stroke. There's some evidence that it can also prompt the birth of new neurons.

It would be unethical to force humans to reduce calories, but some accidental experiments suggest that this might work with us as well. The food shortages in some European countries during the world wars were associated with a decrease in deaths from heart disease, whose rates increased again after the wars ended. The people of Okinawa, practicing their cultural belief called *hara hachi bu,* would eat until they were 80 percent full, routinely consuming 30 percent fewer calories than average Japanese residents. They not only had 35 percent lower death rates from both cardiovascular disease and cancer than the average Japanese population, but until their diet became more Westernized, more residents lived to one hundred than just about any other place on earth.

Also, the eight men and women who participated in Biosphere, an experiment that involved living in a completely closed-off self-sustaining bubble, ended up eating 22 percent fewer calories. As a result, they had, on average, a 17 percent decrease in body weight and marked reductions in risk factors for heart disease, including reduced blood pressure as well as lower levels of glucose and fat.

A recent review of caloric-restriction studies published in the *Journal of the American Medical Association* emphasized that "these associations do not prove causality between decreased calorie intake and increased survival," but added, more optimistically, "these data support the notion that the common link between aging and chronic disease is not inevitable and that it is possible to live longer without experiencing cumulative increase in serious morbidity and disability."

There is now even a Calorie Restriction Society, nine hundred strong and growing, whose members will happily tell you the proper method for weighing arugula and whose recent book, *The CR Way: Using the Secrets of Caloric Restriction for a Longer Healthier Life,* has a handy recipe for a "Delectable Dessert Sandwich," which consists of a piece of bread sprinkled with pumpkin pie spice.

No one knows exactly why reducing calories seems to extend life. It may be that calorie reduction works simply because with less energy burned, there are fewer free radicals produced and less damage to cells. Mattson believes that lowering calories works in other, more important ways as well. A low-calorie diet, he says, puts the body in a state of mild starvation, which in turn puts the body under mild stress and, in turn, activates a continuous stream of helpful repairs.

"The best way to look at this is an analogy with muscle cells," Mattson told me. "Exercise stresses muscle cells with increased energy demand. A lot of free radicals are produced during exercise and the mild stress activates signaling pathways that lead genes to make proteins that protect cells against stress. Intuitively, it makes sense."

And, he added, "The same proteins that are increased in muscle cells in response to stress are also increased in nerve cells in the brain with exercise, cognitive stimulation and dietary-energy restriction."

In particular, Mattson said, mild stress on nerve cells produces a magical repair substance called brain-derived neurotrophic factor, or BDNF, clearly one of the current celebrity substances in aging-brain research. "Over the past ten years," he said, "a huge literature has emerged that this BDNF is important for synaptic plasticity, promoting survival of neurons and instigation of neurogenesis."

Mattson's most recent research was the first to show that monkeys put on a six-month diet with 30 percent fewer calories and given a toxin that destroys dopamine cells in the same way as Parkinson's disease had higher levels of dopamine and better motor functions than monkeys that had the same brain-assaulting toxins and ate regular amounts of food. What's more, those same monkeys had much higher levels of BDNF in their brains.

It remains to be seen whether monkeys with such a reduced diet will live longer. A twenty-year experiment at the National Institute on Aging is still ongoing, though early results are promising. It also remains to be seen if all this could really work in humans over the long run. Most animal diet experiments not only reduce the calories of one set of experimental animals but also *increase* the caloric intake of animals in the control group, which usually gets considerably less exercise as well.

"We know that humans who overeat will not do so well," Mattson said. "The question is, can we take someone who is not overeating, who is normal weight, and, with reducing calories, find that they, too, would have further benefit. We still don't know that for sure. But my guess is that there would be some benefit."

One reason Mattson and others believe in the reduced-calorie idea is that to them it makes evolutionary sense. When food is scarce and the cells sense this through stress, ancient survival mechanisms kick in to protect the organism until food is plentiful again. Those mechanisms include an increase in repair as well as a temporary suspension of reproduction. A severe reduction in calories (think anorexia) shuts down the reproductive system in females because "if there is no food you can't reproduce because there is no food for the children," Mattson pointed out.

Still, severe malnourishment can lead to death. No one knows when a good reduction in calories turns into a bad one that would bring a number of damaging effects. Animals that are fed 50 percent less than would be normal will die. Mattson believes that about two thousand calories for an average male and eighteen hundred for an average female—as long as the diet includes all the necessary nutrients—would likely prevent the body from being harmed and still provide the mild stress necessary to prompt cell protection. At this point, he thinks the idea has real merit.

"Reducing calories activates mild stress that upregulates growth factors that protect the cell against aging and disease," he said.

And this may very well be how antioxidants are really working as well. Mattson thinks it is the toxins in such things as the skin of a red grape (which is there to ward off insects and what gives us the resveratrol) that produce this mild stress that prompts beneficial repairs.

Still, it may not be enough to simply eat piles of antioxidants. To get an impact that way, the dosages would have to be enormous (fifteen thousand glasses of red wine a day, for instance). Such dosage issues may very well be why experiments with supplements generally have not worked in humans.

But we might still get some help from certain foods at more normal levels through the toxic effect. Resveratrol, even at the level of a glass of wine a day, Mattson believes, might be enough to mildly stress cells. (Similar toxins are found in other foods that are antioxidants, such as garlic and broccoli.) And that stress, again, may help boost maintenance systems.

"The benefits in fruits and vegetables might not be because of the antioxidants but because the toxins are producing this mild stress," Mattson said.

In the end, Mattson believes we may eventually crack all this in an easier way. It's true that very few of us will get up in the morning eagerly anticipating that piece of bread with pumpkin pie spice. So he and others are doing their best to isolate the various chemical toxins in plants and, with luck, stuff them or their biological equivalents neatly into a pill. And if that pill can be found, he believes it may be most beneficial in middle age.

"There is evidence to show that exercise, cognitive stimulation, and nutrition can work in middle age," he said.

In fact, a study sponsored by the National Institute on Aging reported in July 2008 that the compound resveratrol slowed age-related deterioration and functional decline of middle-aged mice. Although resveratrol did not make the mice live longer, those that had the substance added to their regular diet, starting in middle age, had lower cholesterol and fewer cataracts, as well as significantly better balance and coordination, than the mice that did not get their dose of resveratrol. Researchers responding to the study, published in the journal *Cell Metabolism*, suggested that resveratrol, naturally found in grapes and nuts, may induce some of the same effects of caloric restriction. As I finished research for this book, scientists were awaiting a similar trial of resveratrol in monkeys.

"Research is attempting to understand the process of aging and to determine how interventions can influence this process," said Richard. J. Hodes, director of the National Institute on Aging. "Dietary restriction has well-documented health benefits in mammals, and the study of possible mimetics of it, such as resveratrol, are of great interest. Resveratrol has produced significant effects in animal models, now including mice, where it mimics some, but not all, consequences of caloric restriction."

Still, it's entirely possible that even after all this, we may simply end up back at the same place we started, proving, as Bickford says, that "the old wives' tales and our mothers were right when they said, simply, 'Eat your fruits and vegetables.' " One of the more ambitious recent studies to address this, by Columbia University's Nick Scarmeas, in fact found that those who ate the so-called Mediterranean diet, heavy on vegetables, had a lower risk of developing mild cognitive impairment over a four-year period, perhaps by improving cholesterol levels, blood-sugar levels, and blood-vessel health overall, or possibly by reducing inflammation.

It certainly makes sense. But we still don't know how, exactly, it might work. After all, even something as simple as the herb thyme, as Michael Pollan, author of *In Defense of Food*, points out, has dozens of antioxidants, with names like terpineol, alanine, anethole, apigenin, ascorbic acid, tryptophan, vanillic acid, selenium, tannin, and on it goes. Which one will prove to be the magic bullet for the brain?

"I don't think we will find one chemical in something like a blueberry that is the active ingredient," Bickford says. "I think we are going to find it is a lot of chemicals working together, that it is more synergistic."

And just as we search for what we should eat, it is just as important to know what we shouldn't. Again, there are very few studies that show that specific foods, like steak, are bad for the brain, but we do know that certain patterns of eating can lead to conditions, such as obesity or type 2 diabetes, that can be harmful. Most of this new research is based on linkage studies and does not necessarily prove cause and effect, but does indicate what we should pay attention to.

There are recent suggestions that type 2 diabetes, the most common kind, which is often related to obesity, may increase the risk for dementia, for instance. It's still unclear if it's the diabetes or the obesity or both that may increase the risk, because not everyone with diabetes gets Alzheimer's and not all those who get Alzheimer's are diabetic.

But in recent years, a number of large studies have found that those with type 2 diabetes are twice as likely to develop Alzheimer's. It may be that the cardiovascular problems caused by diabetes block blood flow to the brain or cause strokes, contributing to dementia. The same kind of plaque that builds up in the brain with Alzheimer's also accumulates in the pancreas with type 2 diabetes. It's also possible that abnormalities of glucose metabolism and insulin levels in the brain may be harmful. Those with type 2 diabetes often have insulin resistance—when their cells cannot use insulin well—so the pancreas makes extra insulin, which builds up in the blood and can lead to inflammation and possible harm to the brain.

One of the new studies, by researchers from the Karolinska Institute in Sweden, found that even people who had borderline diabetes were 70 percent more likely than those with normal blood sugar to develop Alzheimer's. Another study, in Finland, published in the *Archives of Neurology* in 2005, found that being overweight in midlife—even without diabetes—increased the risk of dementia. The researchers looked at the records of 1,449 randomly selected men and women when they were fifty-one and then again when they were seventy-two and found that midlife obesity, like high blood pressure and high cholesterol, doubled the risk for dementia—and that those who had all three risk factors were six times as likely to become demented.

Again, the reasons are unknown, but with a number of studies showing the same findings, it begins to appear that being obese in middle age is not the best thing you can do for your brain. One of the most recent studies, by Scott Small at Columbia, used brain scanners and found a tie between glucose levels and—again—that tiny area of the hippocampus, the dentate gyrus, that is so crucial to memory. Small found that unregulated spikes in glucose were linked to lower blood volume in the brain's dentate gyrus. The effect came with levels of glucose that aren't necessarily seen with diabetes, but with the normal aging process as we reach middle age. And it is known that physical activity and a proper diet—one that leans much more on fruits and vegetables than on highly sugared sodas and snacks—can help regulate blood sugar.

About 20 million people in the United States have type 2 diabetes. The number has doubled in the past two decades and is expected to keep increasing because rates of obesity are rising. Worldwide, diabetes is also increasing, up to 230 million cases from 30 million in the past twenty years.

Obesity rates, too, remain high. According to the Centers for Disease Control and Prevention, 34 percent of U.S. adults aged twenty and over were obese in 2008. Alzheimer's now affects one in ten people over age sixty-five and nearly half the people over eighty-five. About 4.5 million Americans have it, and taking care of them costs $100 billion a year. The number of patients is expected to grow, possibly reaching 11.3 million to 16 million by 2050.

But those projections about dementia do not include a possible increase from obesity-related diabetes. In fact, when scientists talk about the potential for those in middle age now to avoid the destructive deterioration of dementia and retain their high levels of cognitive function into old age, there's often a caveat. That is, if current trends toward increasing obesity continue, then, as one nutritional researcher put it, "all bets are off."

11   The Brain Gym

Toning Up Your Circuits

At first, the task seems as simple as a preschool lesson.

You're sitting in front of a computer and the word *apple* flashes on the screen, then disappears. A minute later, the word *apple* appears again.

The question: Did you *see* that word before?

The next test is just as easy. This time, the computer *says* "apple." Then you hear a string of random words: "lamp," "pen," "dog." Then you hear "apple" again.

The question: Have you *heard* that word before?

No problem.

Now it gets dicier. The word *apple* either flashes on the screen or is read out loud. Then you read or hear a longer list of unrelated words. Then you hear "apple" again.

The question: Did you *hear* "apple" before, as you're hearing it now? Or did you *see* the word? Your instructions are to push a button only if the word *apple* appears in the same form it had earlier.

Oh, dear. You know you've encountered *apple* before. But how? Did you hear it? Did you see it?

You have no idea.

And therein lies one of the most intriguing aspects of our brains. As we get older, most of us, even those in the early stages of dementia, still recognize the familiar. If we look at a word and then see that same word again, we say, "Aha!" Show us *apple*, then show us *apple* again, and, sure enough, we know our apples.

But take it a step further and, starting sometime in middle age, a subset of our memory can grow murkier. We're certain we've come across something—or someone—before, but our recollection of *how* or *when* is lost.

This is that feeling you get at a party—the deep and usually accurate conviction—that you *know* someone, but you don't know where you know him from. Is that guy your daughter's soccer coach? The guy from church? The guy you met with the dog in the park? Did you hear "apple" or see *apple*? This is a task of memory but, again, memory in context that relies on a whole host of brain functions. We have to fire up the proper parts of our brain, recruit more brainpower if we need it, resist the urge to let our minds wander this way and that—and recall what—and how—we have seen or heard something.

Over the past few years—as it has become clear that memory itself is not one thing but many and that some parts age better than others—the question is, can we fix the parts that need a little help? Exercise and food may help. But can we also zero in on our brain's weakest areas and, through training, buck them up?

It is, when you think of it, quite odd. If we have normal healthy brains in middle age, we don't forget that we have a brother in Phoenix or that we once lived in California—basic autobiographical details stay with us. We build richer vocabularies well into old age, proving that even newly acquired knowledge can stick.

We also excel at basic recognition—"Yes, I've seen this word before"—and familiarity, a kind of recognition memory that's so deep-seated we often mistake it for an emotion—"I *feel* like I know that guy."

But other types of memory don't weather as well. Memories for events—how or when something happened—grow hazier. Did I go to cousin Harry's for Thanksgiving last year? Did I buy bread? Did I turn off the stove? Did I *see* or *hear* apple?

Often called episodic memory, this kind of recall is not as automatic as other types of memory. It takes more effort. We have to connect dots, put something in context, like remembering the sequence of a story. It takes a more sophisticated, wide-ranging neural machinery, and, for a variety of reasons, brains, as they age, start to balk.

At middle age, our brains are negotiating the world with finesse, but a few neurons here and there are feeling their years. So can we push those wayward brain cells back in line?

Using More of the Brain

Tricks work. There are a number of strategies aimed at improving overall basic memory functions. Lists are good and may be all some of us need. Studies show that adding contextual detail can help. If you're trying to remember a word, for instance, you can ask yourself if the word is abstract or concrete. If you're trying to register a face, you can focus on a visual detail—a large forehead or a crooked nose. Added information prompts more brain activity in more areas, more connections, and better recall later on.

Indeed, even our imaginations can be helpful. In one fascinating study, neuroscientist Denise Park and a colleague, Linda Liu, found that older people trying to remember to check their blood glucose levels at a certain time did considerably better by first *visualizing* themselves doing that chore. Those who spent three minutes every morning imagining themselves testing their blood sugar were 50 percent more likely to actually do the test later on in the day than those who used other strategies, such as actually practicing the action. They had, through visualization, created what Park calls a stronger "neural footprint" of what they wanted to remember.

Park suspects that using our imaginations may be effective as we age because, again, it relies on a part of our brain's machinery—automatic memory, a more primitive part of memory—that does not decline quickly.

Given that, if we really want to keep our brains in better condition for the long haul, we may need to fundamentally change the way they operate. A small percentage of people may have no problems well into their seventies and eighties (and if that runs in families it could be genetic), but for most of the rest of us who do, we may need to coax our brains—shove them, even—back into younger or more efficient patterns.

At a lab at the University of Toronto, neuroscientist Nicole Anderson is trying to do just that. Using the latest knowledge about how brains work best as they age, she is teaching people to boost their performance by, again, using two sides of their brains instead of one—training them to be bilateral.

The idea here is that those who adapt and learn to use both sides of their brains, or call on their powerful frontal lobes more efficiently when necessary, will be able to stay in better cognitive shape. The questions are: Can those techniques be taught? And can they be taught to those in middle age and beyond in a way that will last?

Anderson is betting that the answer to both questions is yes, and so she is giving it a try. Day after day, men and women file into her Toronto lab and, sitting in front of computers, try to recall whether they've *heard* or *seen* various words, such as *apple, bucket, lamp*. As Anderson explained when I spoke with her before one recent lab session, this exercise specifically targets episodic memory, the ability to remember something in context. How did I encounter that word *apple*? Did I see it? Did I hear it?

To be successful at this, we need to recruit our most elite brain region, our frontal cortex. When we're younger, we use only one side of our brain's frontal areas to deal with such complex contextual information, such as how we encountered the word *apple* or how we know that guy at the party. But as we age, the best and brightest among us start to call on both sides to do this. These brains power up. They become bilateral, using more power to get the job done. Smart brains unconsciously adapt when necessary. They call in the reinforcements.

When I spoke with Anderson, she was in the middle of trying to train a group of adults to make such adjustments. As part of her study, which is ongoing, participants' brains are scanned before the training and afterward. Anderson wants to see if those who initially do not use both sides of their brains learn, as they improve in the task, to do so. Can a brain be taught to use more of itself when necessary?

Obviously, this goes way beyond crossword puzzles. Rather, as Anderson explains, such wholesale brain rehab aims at the underlying processes and trains them.

"We think as people improve they'll start to use more of their brains, that we'll be able to induce a pattern of bilateralization," Anderson said. "We are tapping into something different here. We're trying to train the older brains to do this when they need to, to use their most appropriate and powerful mechanisms."

Video Game Training

If the participants in Anderson's study are explorers on the outer edge of brain enhancement, increasingly they have company. At a lab at Columbia University one recent morning, another brain explorer, Martin Goldblum, settled into a chair to play a video game called Space Fortress, which Columbia neuroscientist Yaakov Stern hopes will train older brains to retain, or return to, more efficient youthful patterns.

An artist, the sixty-six-year-old Goldblum usually doesn't spend his days playing video games. But nevertheless, he was the lab's superstar. As I watched the game, I was surprised that anyone could learn it well, let alone those in middle age or older. But Goldblum had mastered it.

"It's challenging," said Goldblum, a trim, youthful man dressed in jeans and a black T-shirt who was happy to chat as he mouse-clicked along in the game. "You have to multitask; you have to do a lot of different things at the same time—and there is an insistence to it all."

To play, Goldblum grabbed a black joystick with his right hand and a computer mouse with his left. A large green hexagon flashed on the screen and there was a gun that would try to shoot down Goldblum's spaceship, which he had to keep inside the hexagon as he shot down small asteroids that flew by. The game had dozens of rules pertaining to when you were allowed to shoot, how often, and where.

To an onlooker, it seemed impossible. In fact, Yaakov Stern, the cognitive reserve researcher at Columbia University who was running the experiment, was initially unsure if older people could learn the game at all. But learn they have.

And the study had a hidden agenda. One-third of the participants would function as the control group and would not play the game, another third would learn and play the game, and the last group would be told while they were playing the game to "shift focus" and concentrate solely on the activity of the spaceship or, alternately, the asteroids.

What's interesting about this game—and why Stern selected it for his study—is that it has already been shown to work, through a combination of concentration and switching that concentration to train the brain to focus. And, even more important, that increased focus has carried over to the real world. Young military aviators who trained on the same game in Israel did better when they were flying actual airplanes.

The major problem with much of what is sold as brain training is that it has never been shown to work in the real world. People get better at the training game, but that improvement does not necessarily help them remember, say, why they went to the store when running an errand. As Stern says, researchers are now looking to "find an intervention that generalizes, that helps people cope with aging."

At fifty-four, Stern is in the thick of middle age. The hallway outside his office is lined with black file cabinets, data from the dozens of studies he is overseeing. His job involves frequent travel, lectures, students, and, in his spare time, teaching his daughter how to drive. He is, as many of us are in midlife, fully engaged in just about every possible direction. But even as he navigates all that with outward ease, Stern conceded that his own brain now needs more help.

"I have to rely on my Palm Pilot all the time now," he told me. "You know how you used to remember everything really easily? Well, clearly that gets harder. And embarrassing. The other day, I was telling this wise story to the students, and they had to tell me that I had told them that same story the week before."

Stern laughed at this, but he's committed. His family has a long history of Alzheimer's, and he's determined to find a way to combat the disease. In particular, he wants to develop surefire techniques to buffer the brain against the assaults of normal aging and dementia. Stern and many others believe that such a buffer, or brain reserve, can arise innately from a lucky set of genes, but it can also be developed throughout life, from anything that helps the brain stay flexible and strong, such as education or even complex leisure activities, jobs, or perhaps even certain video games.

One hope is that if we could build stronger brains, we could at least push serious mental decline down the road a few years. Statistics show, for instance, that if the onset of Alzheimer's could be delayed for only five years, many of those who have the disease will die of something else, at least saving them from devastating years of dementia.

With his video game study, Stern wants to find a way to boost our brain reserve and make it stick. That means not just perfecting one skill or another, but instead improving how the brain operates overall and coordinates thoughts. Such coordination, often called executive function, goes beyond simple memory and is the specialty—again—of our elite frontal cortex.

It's these coordinating skills that the Space Fortress game sharpens because it forces players to multitask and shift their focus within the context of the game. A number of studies show that if you want to improve certain athletic prowess—say, your tennis game—it's best to focus on one particular skill, but only within the context of a whole game, a test of both focus and coordination.

As Stern explained it, a tennis coach might say, "Okay, today we're going to play but we're going to concentrate on your forehand. We know that kind of training works, and in this game, Space Fortress, when players shift emphasis while they are playing, they are getting that training in attention, focus, and coordinating abilities. And they are perfecting those skills in a real situation, not in isolation.

"A real-world analog to this game would be when you are driving and talking on the phone," said Stern. "As we get older we often don't handle those situations so well, those dual tasks that take attention and focus. We can lose that."

Stern wants to teach older brains to keep that focus, or, if necessary, get it back. And the way to do that, he believes, is to teach them to be more efficient, to more easily tap into their most powerful focus-enhancing frontal cortex. Studies by Stern and others have shown that many of those with high levels of reserve—who appear to be protected from aging or disease a bit better than others—seem to use their brains in just this way.

"It's like a Mercedes and a VW," explained Stern. "You have to push down on the accelerator more to get that VW going, and when you get out on the autobahn, an old VW will peter out at seventy miles per hour. But the Mercedes will start up with less of a push and keep whizzing along easily. Those people with cognitive reserve are like Mercedeses. Their brains are more efficient. And it's not just IQ. It's not just something you are born with. It can come from lots of things."

The question is, can we find specific ways to build this ability. Can we take a VW brain at middle age, when Stern and others think we may have to start all this, and morph it into a Mercedes brain—with a video game? Or can we take a Mercedes that's a bit past its prime and return it to its autobahn days?

"We don't know yet but I hope so," Stern told me, adding that the only way to find out is to do long-term, serious studies to figure out what stimulates a brain in a way that makes it stronger, to "find out what really works and how."

Plastic, Mutable Brains

When we're thinking about how to keep a brain on track as we age, however, there's a bit of a rub: We're all pretty stimulated already. We live and work in complex, multitask worlds, with CNN blasting, and economic crises to worry about. We're *already* quite enriched.

Can we find any add-on training program that has a prayer of busting past the bustle of our daily lives and pushing our brains even further to maintain or return to better habits?

At a recent meeting of cognitive scientists, Michael Merzenich, who has developed a new brain-training system of his own and is one of the leading proponents of this brain-training idea, fairly bellowed at his colleagues. "I can improve anyone," he said, his face growing red as he spoke at the cognitive scientists' meeting in Washington, D.C. "Anyone!"

Now a professor emeritus at the University of California at San Francisco, Merzenich is not one to be taken lightly. One of the pioneers of neuroplasticity, he has often been ahead of the curve. Nearly thirty years ago, he showed how a monkey brain reassembles itself after injury. A member of the prestigious National Academy of Sciences, he invented a system that uses sounds to help some dyslexics. He holds fifty patents.

And he has now moved on to the older brain, which he *insists* can be trained to act like a more efficient brain if only we would try hard enough. Several years ago, Merzenich started a company, Posit Science, and developed a system that he says can push the brain, not just to adapt, but, in areas where we experience loss, to get it back to where it was years earlier. He believes he can turn back the clock in our brains.

At middle age, our brains have learned to handle our world with ease—and the hope is that we can keep them at that level as we age. Merzenich, for one, thinks it's possible.

"Strategies and compensation are good but I don't think you have to just correct," he said when I spoke with him recently. "You are not stuck with the deficits you have or the negative changes that occur. You can fix them."

One might be skeptical about such definitive pronouncements, even from someone as persuasive as Michael Merzenich, but he does have some evidence to back him up. In 2007, Elizabeth Zelinski of the University of Southern California announced results of the first official study of Merzenich's computer-based training system called Brain Fitness. The study, officially published in 2009, was a kind of randomized, double-blind trial that is the gold standard for evidence. And it found that Merzenich's form of intense brain training appeared to work.

Performed at centers all over the country, including the Mayo Clinic and Veterans Administration hospitals, the study was based on five hundred adults who were divided into two groups. One group trained an hour a day for eight weeks on Brain Fitness, which meant that the participants sat with headphones on in front of a computer and did a series of exercises set up to fine-tune their brains. The training was largely auditory in that they had to either distinguish between similar-sounding words, like *mat, pat,* and *cat,* or decide if a sound was whooshing up or down. The other group of adults, the control group, spent the same amount of time watching educational DVDs. The study was funded by Posit Science Corporation, but none of the investigators received money from the company.

At the end, those who had the computer training performed better; in fact, they performed like people ten years younger on standard cognitive tests. Those who only watched the movies did not improve.

Over the past few years, we have been inundated with computer-based brain-training games of all sorts, from Nintendo's Brain Age to Happy Neuron. The industry has ballooned from $2 million in 2002 to $80 million in 2007. But there's scant evidence that these commercially available games have any concrete effect on our brains at all.

With the Brain Fitness training, at least we have a commercially available program that has been tested. "I started out as a skeptic. I really didn't believe it would work," Zelinski told me when I spoke with her. "I think in the end it came out much better than I ever expected."

Clearer Signals, Better Focus

Brain Fitness targets several key brain functions. The first is fidelity, or the clarity of the signal that initially enters the brain. The theory is that by learning to better discriminate between similar sounds, the older brain is forced back into the sharper patterns of focusing that it had when it was younger.

"The idea behind this is that there is a dark side to the brain's plasticity, and that as we age we can stop paying attention, we can stop focusing," Zelinksi, who was a lead author on the study, said. According to Merzenich's idea, information coming into the brain as it ages can get fuzzier, not at the level of the actual ear, which can have its own problems, of course, but along pathways leading from the ear to inner-brain regions. If the incoming signals are "noisy," then the information that's stored and copied in the brain will also be noisier, more chaotic, less useful. On the other hand, if you can retrain the brain to focus and get the information and the signals sharper to begin with, you will get a more durable memory.

Additionally, the more distinct signals will stimulate the brain to produce the right kind of brain chemicals, the neuromodulators such as dopamine, norepinephrine, and acetylcholine, that help us to learn and consolidate memories—the same ones that tend to decrease with age.

"I tried this training myself and it was very, very difficult," Zelinski told me. "It forces you to be engaged, to pay attention. But it's not so hard that you give up and lose focus. This approach really bludgeons the brain to work hard."

Just when such bludgeoning should begin remains unclear, however. Certainly, what happens to our brain at every age matters, even in the womb. But if we start this neural training at age fifty, say, will it really help us to both find and tie our shoes when we're eighty-five?

Most believe that middle age is a prime time for all this, but that has yet to be proven.

"There may be a timing issue with all this. We just don't know yet," said Zelinski.

Merzenich, for one, insists that maintaining or pushing a brain to adopt better habits is a "realistic goal," and he is convinced we should start down that path before we are too old.

"At middle age we are pretty good at manipulating the information that is coming in. Your brain might not be as fast as when you were twenty, but you have twenty or thirty more years of experience at manipulating information so that the brain can do it pretty efficiently. At middle age, that experience trumps the declines and your brain is operating pretty well for you," Merzenich said. "But if we are in a job where having a good brain matters, we want to keep it operating at that level. At some age, there is a tipping point, where experience no longer trumps the losses in the brain. That tipping point for most now is probably sometime in our seventh decade. But we want to improve that and, if we can, change the slope of the trajectory."

To change that slope, though, may take elaborate training and hard work. As we age, we tend to fall into predictable patterns with life and brain activity, as Merzenich says, on "autopilot."

"It's not going to work if we keep doing the same thing over and over again," Merzenich told me. "As we age, we fall into behaviors that are more and more stereotypical and more limited. We are not working as intensely at refining or maintaining the high level of operations of our brains. It's not just acquiring new information. There's nothing wrong with that, but it's not enough."

To keep a brain from becoming lazy, we might even need training tailored to each brain function, much as we go to a gym and work on our thigh muscles one day and our triceps the next, because, as

Merzenich said, "there is no one magic bullet" for the brain and "one Sudoku puzzle" will not be sufficient.

Instead, we have to deal with the aging brain from a number of angles. We have to force ourselves to pay attention—to concentrate at a level of intensity that is, as Merzenich put it, "the stuff of childhood." And we have to get out of our ruts.

"People have to understand that there is a direct relationship between the physical growth of the brain and how you use it," Merzenich said, "and as the idea sinks in that we can drive positive changes in the brain, it is empowering."

Beyond Crossword Puzzles

At the moment, though, it's still extremely hard to prove that any of the brain-training programs work outside the lab and carry over into real life. In the study of Merzenich's system, in fact, there was only a hint that this was true. Those who completed the training said they "perceived" that they were better at remembering phone numbers and names in their actual lives. Those who had simply watched educational DVDs reported no such improvement. Those who completed the training also had higher levels of those helpful growth factors in their blood.

"But we just don't know," conceded Zelinski. "We just don't have good data on how this translates into the real world. You can't really follow people around and see if they're forgetting stuff."

Still, we know enough now to at least get started or think about getting started. Zelinski fervently believes we should get going now—and take the tougher road.

"Crossword puzzles are not enough. You are mostly trying to find words you already know," she said. "We need to challenge our brains. It should be something hard for you—not so hard that you lose focus—but hard. We need to get out of our comfort zones."

Concerned herself, Zelinski is doing whatever she can to maintain the power of her fifty-five-year-old brain. She is taking piano lessons.

"My son is eighteen and he's been playing since he was in third grade and he can read music and beautiful things come out of his fingers. I wanted to be able to do that. I had this longing," she said. "So when I had a sabbatical, I started taking lessons. It's very slow and hard and to do it you have to find a teacher who will work with you. But, hey, they work with five-year-olds."

One small study showed that a group of sixty-year-olds who trained themselves on piano were, after six months, better on cognitive tests. Zelinksi readily admits that such a tiny and "cute" study does not equate with firm evidence. We still have no idea how many sonatas or sound whooshes it takes before our lazy neurons wake up, or if we have to do it in some particular order, or if it will really make a difference in the end.

Still, how can it hurt? Most of us with memory issues are not headed for dementia, but since dementia begins in the brain long before it's evident, how can we be sure? As Zelinski pointed out, "It will be thirty years until we know," so we might as well "prepare for the worst and start moving."

Emotions and Cognition

Along the way, we might as well try to enjoy ourselves, too, because our moods are also surprisingly important to our brains. Boxes of studies have found that people who are less grumpy, less lonely, happily married, or otherwise entwined with their fellow human beings or even their pet beagle have a lower risk of developing heart disease or Alzheimer's, a better chance at staying mentally alert, and a greater likelihood of a long life. One recent study in England found that people in middle age who simply popped down the street to their neighborhood pub on a regular basis had better cognitive skills than their sit-at-home neighbors.

It's true that many of these mood-brain-health connections have, through the years, also had the classic chicken-and-egg problem. It's entirely possible that happy, optimistic, pub-hopping sorts are a special breed to begin with, perhaps much more likely to take other steps that improve their health, such as taking their blood pressure pill on time or eating their carrots. It's also possible that those who are more social are simply healthier overall, with no secret disease cooking away to make them cranky.

Lately, though, more controlled studies have gotten a much better handle on the question. These are true comparison studies that have one group doing something specific, say, joining a choir, and another sitting at home. Then the two groups are compared in a serious, head-to-head way. And those studies, which try to eliminate other possible influences, such as gender, obesity, smoking, and levels of education, now routinely report exactly the same thing that the earlier softer studies found: Being sociable and cheery is good for both body and mind.

One recent study by researchers at Johns Hopkins University, for instance, found that men and women who volunteered to tutor students through a group called the Experience Corps at Baltimore schools had a slower rate of decline in their memories than those who were put in the nonvolunteer control group.

And that was the case even though both groups started out with similar levels of cognitive function. The volunteers also significantly decreased TV watching at home and reported that they felt much stronger physically (the students did better in school, as well).

My own personal favorite of this current rash of "be-social" studies was a recent one by researchers at the medical school at the University of Miami that looked at how architecture might improve the brains of sixteen thousand older residents living on one block in the Little Havana section of Miami. The study found that those who simply lived in houses or apartments with balconies that faced the street and encouraged neighborly chattiness had better cognitive function than those who did not have such architectural benefits.

Even self-image may matter. A study by Becca Levy, a psychologist at Yale University, found that the memories of older people improved after simply seeing positive words about aging. In Levy's study, the words were flashed too quickly for people to be aware that they had read them, but nevertheless, on some level they had an effect. People did poorly if they first saw negative words, such as *decline, senile, decrepit, dementia,* and *confused.* But memories significantly improved if they were first exposed to positive words about aging, such as *wise, alert, sage,* and *learned.*

Similarly, Thomas Hess, the psychologist at North Carolina State University, found that attitudes are self-fulfilling. In his studies, older people did worse on memory tests if they were first told something negative about growing old, such as that the upcoming study was on how aging affects learning and memory. But if they were first told something positive, such as that there was not much of a decline in memory with age, their memories on the tests improved. Another recent study, which keyed in on our competitive natures, found that people in middle age and older did better on cognitive tests if they were told they were being tested with younger rather than older people.

It's not known precisely how self-image, rich social connections, or peppier moods affect the brain. But there are some good—and fascinating—hints.

One candidate is stress. If social interactions can ease stress—and that means you have to pick your friends carefully, of course—the brain benefits. Unrelenting stress, in particular sustained levels of stress-induced cortisol, kills neurons in the memory-rich hippocampus. Depression, too, has been linked with a smaller hippocampus.

And there's emerging evidence that our brains are set up from the get-go to cooperate, so they may work better if we do. One brain-scanning study by the National Institute of Neurological Disorders and Stroke found that when participants made a choice to share some money they were given rather than keep it for themselves, the brain's reward center became active, the same dopamine-releasing area that comes alive with sex, chocolate cake, and cocaine.

Mirror, Mirror in the Brain

And then there are mirror neurons. Over the past few years, neuroscientists have discovered an entire new class of brain cells that appear to exist primarily to help us recognize and feel the joy and pain of others. These mirror neurons may be one of the strongest neurobiological underpinnings of our drive to connect—and yet another reason why we have to be fairly careful with whom we are connecting. Mirror neurons make emotions contagious.

Even the story of the discovery of mirror neurons is fascinating. About twenty years ago, a team of Italian neuroscientists went out for lunch and left a research monkey hooked up to electrodes. When they returned and walked into the lab, one of the researchers lifted an ice cream cone to his mouth. As he did this, bells and lights started going off, indicating that the areas of the monkey's brain that corresponded with lifting a hand to a mouth to eat something also became active. The monkey was not eating an ice cream cone, but in his brain he was.

From that surprise finding, the field of mirror neurons exploded. It's now thought that mirror neurons, which are scattered in pockets all over our brains, are what help us understand the motives and actions of others. It is our mirror neurons that make us "feel" the pain of the character in the movie who's being dumped by his girlfriend or the fear of a kidnap victim with a gun to her head.

"When we watch the movie stars kiss on screen some of the cells firing in our brains are the same ones that fire when we kiss our lovers. . . ." writes Marco Iacoboni, a neurologist at UCLA, in *Mirroring People.*

"When we see someone else suffering or in pain, mirror neurons help us to read his or her facial expression and actually make us feel the suffering or the pain of the other person. These moments . . . are the foundation of empathy and possibly of morality, a morality that is deeply rooted in our biology."

Sociability as Exercise

And there's increasing evidence that being with other humans helps tone our brains' dendrites as well. Being social is hard and complicated and it taxes the brain.

"People forget how difficult and complex a task social interaction is," says Denise Park, the neuroscientist in Dallas. "There are a lot of demands just in meeting new people. You have a name and face and you have to integrate that name with that face. And that new person will have personal history they're telling you. And you will be telling them things about yourself. And the next day when you see them again you have to bring all that back to mind. Social engagement, sustained social engagement, is cognitively very demanding."

When I spoke with her, Park had just finished the pilot phase of a new study to test this further. In the study, socially isolated middle-aged and older adults come to a center to learn digital photography and quilting. While learning these new skills will help, Park expects that a large part of the benefit will arise from the human interaction itself.

After the initial part of the study, Park will scan the brains of participants to see if there is increased brain volume in areas associated with learning complex tasks such as quilting or photography. But she expects to see the impact of being socially engaged as well.

As we've said, the brain as it ages has a tendency to wander, to get distracted. Park believes we need to find ways to keep the brain from slipping into its disengaged default mode. She, too, thinks we have to retrain our brains to get back to younger habits. And it's possible that simply being socially engaged will help keep that default mode at bay. Something as seemingly simple as chatting with a friend may push a brain out of its daydreaming tendencies and instead activate powerful, focus-tuning frontal regions. Park and a number of others now believe that we need to—and can—find ways to jump-start our brains *out* of that default mode, teaching them to "call in the troops, the frontal lobes, when they need them."

But we also have to remember that science, try as it might, can't study everything. No doubt we already do a fair amount of good things for our brains; we just don't appreciate them enough—or give them their due.

As I mentioned earlier, after a brain science conference in Washington, D.C., I had dinner and then shared a cab with scientists Denise Park and Laura Carstensen. In the cab, Carstensen told a funny story about how she had given her two-year-old grandson a present that she'd bought at a gag store, a duck that laid an egg. She said she had shown him how it worked, squeezing the duck to push out an egg, but he didn't get it, he was too young. In fact, she said, he was "horrified." Carstensen's attempt to amuse her grandson had laid an egg.

It was a silly story, but—well, we'd all had some wine—it made all three of us laugh. And as Carstensen continued to tell the story and we continued to laugh, the cab driver, an elderly man whose gray hair peaked out from under a worn watch cap, turned around and grinned.

"You know," he said, "you all have the secret of life. Laughing is the secret. You will never get old."

Wisdom from cab drivers is one of the oldest clichés around, of course. But if we know anything in middle age, it's that many of the world's clichés turn out to be true.

Epilogue

A New Place for Better, Longer Lives

So what, then, shall we do with our indefatigable, our inestimable—our newly appreciated—middle-aged brains?

Perhaps it's time for a middle-age revolution.

After all, we have numbers on our side. For all of our planet's history, children have outnumbered grown-ups. A huge pile of the young has been the base of a population pyramid that tapered off toward the top, with progressively dwindling numbers of older people.

But that has already changed. More than 500 million people worldwide are sixty-five and older. By 2030, one in every eight people on earth will be middle-aged or older. For the first time in history—and possibly for the rest of human history—people over the age of sixty-five will outnumber those under the age of five.

Not surprisingly, there's a fair amount of agitated hand wringing about all of this. And it's not just about keeping Social Security funded. Already some believe that any planet filled with so many older humans is a planet in peril, brimming with brains that have already begun to slide.

But what if that peril can be averted. If we keep pushing down our blood pressure, sidestepping strokes, taking that brisk walk, many of us could remain in one fairly well-oiled piece. With emerging evidence that our middle-aged brains are already considerably better than similar middle-aged brains were twenty years ago, and as education levels and wealth continue to climb, there's more than a fighting chance that this good-news trajectory will continue.

But before your endorphins surge with that good news, we have to pause here, too.

It's also true that even if we get half of all that right, there are hidden traps.

Imagine a planet where most of us are in our sixties, seventies, or eighties—or even older. Our hearts are hearty, our bones are sturdy, our brains are blooming.

What exactly are we going to be doing?

There may be some, even in such robust condition, who will be content to swing in *hamacas* in Baja. But most of us will want a little something to do. As Laura Carstensen of Stanford University says, quite bluntly and doubtless correctly: People want to work.

And that's not going so well. According to a 2007 survey by the consulting firm McKinsey & Company, nearly half of baby boomers say they expected to work past sixty-five but only 13 percent of current retirees actually worked until that age in the group that was studied. Rather, 40 percent were forced to stop working earlier than they'd hoped. The average age when they stopped getting a paycheck: fifty-nine.

Age discrimination is alive and well. Researcher Joanna Lahey recently sent out four thousand résumés to firms in Boston and St. Petersburg, Florida, and found that a younger worker was more than 40 percent more likely to be called in for an interview than a worker fifty years or older.

We've extended our lives by dozens of years—and we're finally finding tantalizing new ways to extend our brain spans as well. But we have not taken a nanosecond to think about what to do with all those better years and better brains.

We need a new plan. Maybe we could find some way to move the furniture of our lives around a little bit here and there. Perhaps it's time to acknowledge that what worked well for a vastly simpler agrarian society of the nineteenth century might not be the best fit for the demographically shifting realities of the twenty-first.

We could set up a world that makes sense for current life spans, with more flexible time to raise kids and work during the beginning and the middle and less down time later on.

"It's very frustrating to me because we added thirty years to life and we have put it all at the end," Laura Carstensen told me when I ate lunch with her at Stanford. "Whenever you ask Americans what is their biggest problem, they say time. We love work, but we love our families, too. And we have constructed a life course in which we have to do all of it at once.

"There is no reason it has to be this way. We need to build sabbaticals in all occupations so we can go learn something new. And we need to think about ways to reduce the workweek to three or four days when we have families. At that point in life, we all need to have jobs that are more flexible. And we need to not be irrelevant at age sixty-five."

All that sounds good to me. Is it just wishful thinking?

Not necessarily. As we live longer and have fewer children, we face—indeed, we already have in some places—a severe labor shortage despite the vagaries of economic cycles. If such trends continue, the world will change because it has to, driven by basic economics.

In a few places, this idea has already begun to sink in. Admittedly, the movement is minuscule, but it's out there. Some worried countries have set up ways for older workers to stay in the workforce and help younger workers learn their skills. Finland, which, like many other European countries, now has both a low birthrate and a rapidly aging population, has raised the retirement age. Some companies there now have master programs, in which workers past sixty train younger workers.

Encouraged by studies that show that older employees, contrary to stereotypes, often cost less than younger ones because they take less sick time and are cheaper to train, some companies in this country are waking up, too. Awhile back, Home Depot and Borders started hiring retired workers on purpose, giving them perks such as flexible and part-time hours and, in some cases, the ability to split job locations, working in cold areas in the summer and warm areas in the winter.

Other companies have said that they may resort to filming older workers doing their jobs so that knowledge about how things actually work won't just walk out the door. One energy company that runs nuclear power plants recently asked managers to identify the "degree of criticality of knowledge" of older employees so that they can transfer at least a smidgen of that criticality to younger brains before it's too late.

We live in a strangely schizophrenic world in terms of age. We tell people to get out of the way at sixty-two—too old to teach, too old to be a doctor, too old to be a lawyer—yet we've had a man running for U.S. president, arguably the toughest job around, at age seventy-two. We send clear messages to women, in particular, that they're past their prime in dozens of ways by their late fifties, and yet we have had a grandmother, at age sixty-eight, running the U.S. House of Representatives.

What exactly is too old anyhow? When, in terms of our bodies and our brain cells, are we over the hill?

Stanford University economist John B. Shoven recently came up with an entirely new way of calculating when we reach the crest of that hill. Given the fact that we're all in better shape and living longer, he argues that our true age should be determined not by years since birth but by years left to live. In this way, he has reconfigured the traditional arc of our lives to create a long period of youth followed by shorter periods of middle age and old age. That means that if you have less than a 1 percent risk of dying within a year you can consider yourself young, and you're not old until you have a 4 percent chance of dying within a year.

Middle age, according to this marvelous system, would be defined by a mortality risk between 1 and 4, a span of time that, by Shoven's analysis of 2000 U.S. Census data, now begins for men around age fifty-eight and for women at age sixty-three.

And under this interpretation, men don't become officially old until age seventy-three and women don't cross that line until they're seventy-eight.

If you take his message to heart—and why not—it means that the coming avalanche of aging brains will not be a disaster. We'll simply turn into a planet of astonishingly competent grown-up brains!

So while we have time and those still-functioning grown-up brains, we might want to prepare a bit, perhaps ignite a tiny insurrection.

The best way to start, to my mind, is to finally give our middle-aged brains the respect they deserve. Maybe it's time to take that broader middle-aged view—to appreciate and put to fuller use—what we still have inside our heads.

We all talk about the benefits of experience, but we forget that all that experience isn't built up in our knees but in our brains. Knees come and go—we can even have them replaced—but we hang on to our brains. And those brains—silent, hidden—have been quite busy building up the rich connections that let us know what we need to know about our world.

By middle age, our brains have trillions of carefully constructed links and pathways that make us smarter, calmer, wiser, happier.

These are the connections that let us, in an instant, recognize the underlying patterns around us and make sound judgments—good choice, bad choice, friend or foe? By middle age, our brains navigate complex situations and complex fellow humans almost on autopilot. Our middle-aged brains simply know that the deal for the latest video-conferencing cell phone/life organizer is no deal at all, know that we don't have to panic because our daughter's latest oddball boyfriend won't last long in the end, know that it really is better to keep our mouths shut if, in fact, we have nothing useful to say, and know when we must speak up to make a difference.

One woman, sixty-two, a writer, summed up all this recently. She told me that she can no longer remember details as well as she did even a few years ago.

"I'm reading a six-hundred-page book now on race relations and it's really good and even a few years ago I would have been able to keep the whole book, all the dates, in my head easily as I read along. And that's just not the case now; my head is like a sieve for facts."

Then she added: "But it's also true that I almost never come across a life problem, domestic or professional, when I don't know what to do, what to say. I feel like I can handle almost any crisis. And if that's the middle-aged brain, giving up facts for solutions, well, to me that's a good trade-off."

As I completed research for this book, a real-life example of this middle-aged solution-oriented brain at work occurred that could not have been more dramatic. It was when the fifty-seven-year-old pilot Chesley B. Sullenberger III ditched his US Airways jet in the Hudson River after a flock of geese flew into the plane's engines, shutting them down shortly after takeoff from LaGuardia Airport in New York. This white-haired pilot, calling on all the established patterns and connections built up in his brain over the years (including, luckily, experience from working as a glider pilot), decided to avoid densely populated areas and set his plane down in the icy river, in a landing that was so controlled, the jetliner stayed intact and all 150 passengers survived.

In this instance, not only the pilot but the entire crew of the plane was middle-aged, or "senior," as they were called, as were the tugboat and ferryboat captains who decided to turn their boats around and speed to the rescue. Middle-aged pilot, middle-aged crew, middle-aged boat captains—they all did the right thing automatically. This was the kind of stark reminder that every now and then elicits widespread appreciation, albeit temporary, for the calm and competent older brain.

"The pilot," said one news report, "handled the emergency landing with aplomb and avoided major injuries, evacuating the plane . . . calmly and in the middle of the river." One reason everyone survived, an air safety investigator concluded, was that the crew were all older, a "testament to experience."

Some see signs that we are edging toward a new appreciation for such experience—one way or another. Writer Gail Collins recently noted that the 2009 Best in Show winner of the prestigious Westminster Kennel Club dog show was a ten-year-old Sussex spaniel named Stump—the oldest dog to win the title in the show's 133-year history—the actor du jour that same year was Mickey Rourke, who was fifty-six, and Mick Jagger was still touring—all signs, she concluded, that "old is in."

"Is this a baby-boomer plot?" she asked, only half kidding. "My own personal theory is that we're witnessing a defense mechanism triggered by the current economic unpleasantness. Since it appears that nobody is ever going to be able to afford to retire, we're moving into an era in which having your car fixed or your tonsils removed by a seventy-five-year-old will need to seem normal. . . . So it's better if we readjust our thinking and start regarding everybody as

20-years-younger than the calendar suggests. . . . Then you will feel much better when the 80-year-old postman delivers your mail and it includes a request for money from your 38-year-old offspring doing post-post-post doctoral work at Ohio State."

Well, maybe we aren't quite there yet, but perhaps it's time for a deep breath and time to get our grown-up brains in gear.

There's stuff to do.

That's not to say, of course, that there won't still be—moments. We have to be prepared to make a few adjustments as well. There will be days when we run smack into our middle-aged brains. I don't want to paint an overly rosy picture.

Which leads me to my favorite middle-aged brain story. Not long ago, an old friend—the owner of a well-accomplished, highly capable middle-aged brain—stopped on her way home to buy actual roses.

"They were beautiful," she said as she told me the story.

"I came home and put them in a vase in the living room."

A few hours later, busy in the bedroom of her tiny Manhattan apartment, she thought: What's that smell?

She thought someone had sprayed perfume around somewhere.

Then, walking into her living room, she suddenly felt thoroughly, utterly ridiculous. There, sitting on the table, was the vase of roses—tall, beautiful roses.

"I didn't even remember buying them," she said.

Not a good story, you say? Well, hold on. Let that wise amygdala do its work and take the more expansive, positive, optimistic view.

After all, thanks to her middle-aged brain, my friend got to enjoy the first wonderful whiff of those tall wonderful roses not just once, but twice.

Acknowledgments

My middle-aged brain had a lot of help with this book largely from other middle-aged brains.

I am especially grateful to the many scientists I write about here. Laura Carstensen of Stanford University, in particular, generously shared not only her science but her enthusiasm and guidance. Others helped more than I could have wished for, too, including Denise Park, Mara Mather, Sherry Willis, Adam Gazzaley, Susan Turk Charles, Elizabeth Zelinski, Neil Charness, John Gabrieli, Deborah Burke, Nick Fox, Mark Moss, Scott Small, Yaakov Stern, Bruce Yankner, Thomas Hess, George Bartzokis, Naftali Raz, Nicole Anderson, Cheryl Grady, Paula Bickford, Joe Mikels, Daniel Mroczek, Valerie Reyna, Christopher Soto, Karen Fingerman, Carol Ryff, Victoria Bedford, Michael Merzenich, Arthur Kramer, and Fred Gage.

The staff at the National Institute on Aging, as well as its director Richard Hodes, never failed to steer me in the right direction.

I am also enormously grateful for all the help I got from friends who are themselves scientists—and middle-aged scientists at that—including Abigail Zuger, Karen Pennar, Barbara and Tim Pedley, Nella Shapiro, and Meg Cameron. I got help, support, and an endless stream of real-time examples from a large group of very smart middle-aged friends, too, in particular Connie Rosenblum, Michael Powell, Evelyn Intondi, Kristen Kelch, Frank Spain, Catherine Woodard, Jack Schwartz, Gail Fell, Phillis Leftin, Bill Walsh, Caroline Grist, and Charles Tremayne. I am also lucky to work in the Science Department of the *New York Times*, whose standards have been so high for so long that they can't help but rub off somehow. In particular and as usual I got a lot of good advice from Gina Kolata, Erica Goode, and Denise Grady.

I got research help from Stephen Sinon, as well as Chris Goelitz, who is not only wise herself but did a lot of the early digging into the topic of wisdom.

The direction and guidance from my editor Wendy Wolf was invaluable, as was the continued support in all sorts of areas from my agent Katinka Matson, whose assistant Karla Taylor also helped.

And, then there's my family. To Richard, thanks for everything. And thanks, too, to my daughters, Hayley and Meryl, who while not yet middle aged, are already wise in their own ways.

Sources

This book is based primarily on extensive interviews with dozens of neuro-scientists, psychologists, and cognitive researchers either in person, on the phone, through e-mail, or at conferences on the aging brain. When appropriate, I have also cited the principal scientific studies whose findings form the framework of the book. I relied, too, on a number of excellent books on aging and the brain, as well as interviews with just about anyone with a middle-aged brain who agreed to talk with me. The following gives sources by chapter.

Introduction

Ephron, Nora. "Who Are You?" Op-Ed, *New York Times*, August 12, 2007.

Safire, William. "The Way We Live Now: On Language; Halfway Humanity." *The New York Times Magazine*, May 6, 2007, 32.

Patchett, Ann. "Mind Over Matter." *Real Simple*, September 2007, 83.

2 The Best Brains of Our Lives

Much of the beginning of this chapter comes from interviews with and the work of Sherry L. Willis, professor of human development at the Gerontology Center at Penn State University in State College, Pennsylvania. Willis and her husband, K. Warner Schaie, have for many years run the Seattle Longitudinal Study, whose ongoing results are reported in dozens of scientific studies, books, and articles.

Following are the books I found particularly useful:

Willis, Sherry, and Mike Martin, eds. *Middle Adulthood*. Thousand Oaks, CA: Sage Publications, Inc., 2005.

Willis, Sherry, and James D. Reid, eds. *Life in the Middle*. San Diego, CA: Academic Press, 1999.

Willis, Sherry, and Susan Whitbourne, eds. *The Baby Boomers Grow Up*. Mahwah, NJ: Lawrence Erlbaum Associates, Inc., 2006.

The cognitive abilities quiz is from a sample test from the Adult Mental Abilities Test Word Series. (Adapted by special permission of Consulting Psychological Press, Inc., Palo Alto, California. From Schaie-Thurstone Primary Mental Abilities Test, 1985. Constructed by Judith Gonda, 1978.)

Other major studies and book segments on which parts of this chapter were based:

Schaie, K. W., S. L. Willis, and I. L. Caskie. "The Seattle Longitudinal Study, Relationship between Personality and Cognition." *Aging, Neuropsychology, and Cognition* 11 (2004): 304.

Birren, J. E., and K. W. Shaie, eds. "Intellectual Development in Adulthood." *Handbook of the Psychology of Aging*, 3rd ed. San Diego: Academic Press, 1996, 291–319.

Willis, S. L., K. W. Schaie, and A. O'Hanlon. "Perceived Intellectual Performance Change over Seven Years." *Journal of Gerontology: Psychological Sciences* 49 (1994): 108–18.

Willis, Sherry L., Sharon L. Tennstedt, Michael Marsiske, Karlene Ball, Jeffrey Elias, Kathy Mann Koepke, John N. Morris, George W. Rebok, Frederick W. Unverzagt, Anne M. Stoddard, and Elizabeth Wright (for the ACTIVE study Group). "Long-term Effects of Cognitive Training on Everyday Functional Outcomes in Older Adults." *Journal of the American Medical Association* 296, no. 23 (2006): 2805–14.

Zelinski, Elizabeth, and Robert F. Kennison. "Not Your Parents' Test Scores: Cohort Reduces Psychometric Aging Effects." *Psychology and Aging* 22, no. 3 (2007): 546–57.

Krampe, Ralf, and Neil Charness. "Aging and Expertise." Chap. 40 in *The Cambridge Handbook of Expertise and Expert Performance*. New York: Cambridge University Press, 2006, 723–42.

Charness, N., M. Tuffiash, R. Krampe, E. M. Reingold, and E. Vasyukova. "The Role of Deliberate Practice in Chess Expertise." *Applied Cognitive Psychology* 19 (2005): 151–65.

Salthouse, T. A. "The Processing Speed Theory of Adult Age Difference in Cognition." *Psychology Review* 103 (1996): 403–28.

Taylor, Joy L., Art Noda, and Jerome A. Yesavage. "Pilot Age and Expertise Predict Flight Simulator Performance." *Neurology* 68 (February 2007): 648–54.

University of Michigan, Press Release. http://www.med.umich.edu/opm/newspage/2008/hmcognitive.html.

3 A Brighter Place

The beginning of this chapter draws primarily on interviews with and the work of Laura Carstensen, Mara Mather, Susan Turk Charles, Joe Mikels, and John Gabrieli.

Among the principal studies I referred to:

Mather, Mara, Turhan Canli, Tammy English, Sue Whitfield, Peter Wais, Kevin Ochsner, John D. E. Gabrieli, and Laura L. Carstensen. "Amygdala Responses to Emotionally Valenced Stimuli in Older and Younger Adults." *Psychological Science* 15 (2004): 259–63.

Charles, Susan Turk, Mara Mather, and Laura L. Carstensen. "Aging and Emotional Memory: The Forgettable Nature of Negative Images for Older Adults." *Journal of Experimental Psychology: General* 132 (2003): 310–24.

Carstensen, Laura, and Joseph A. Mikels. "At the Intersection of Emotions and Cognition: Aging and the Positivity Effect." *Current Directions in Psychological Science* 14, no. 3 (2005): 117–20.

Carstensen, L. L., and B. L. Fredrickson. "Influence of HIV Status and Age on Cognition Representations of Others." *Health Psychology* 17 (1998): 494–503.

Charles, S. T., M. Mather, and L. L. Carstensen. "Focusing on the Positive: Age Difference in Memory for Positive and Negative and Neutral Stimuli." *Journal of Experimental Psychology* 85 (2003): 163–78.

Carstensen, L. L., H. H. Fung, and S. T. Charles. "Socioemotional Selectivity Theory and Regulation of Emotion in the Second Half of Life." *Motivation and Emotion* 27 (2003): 103–23.

Mather, M., and M. R. Knight. "Angry Faces Get Noticed Quickly. Aging Does Not Impair Threat Detection." *Journal of Gerontology: Psychological Sciences* 61 (2006): 54–57.

Mather, M., and L. Carstensen. "Aging and Attentional Biases for Emotional Faces." *Psychological Sciences* 14 (2003): 409–15.

Mather, Mara, and Marisa Knight. "Goal Directed Memory: The Role of Cognition Control in Older Adults' Emotional Memory." *Psychology and Aging* 20, no. 4 (2005): 554–70.

Mikels, Joseph A., Gregory R. Larkin, Patricia A. Reuter-Lorenz, Laura Carstensen. "Divergent Trajectories in the Aging Mind: Changes in Working Memory for Affective Versus Visual Information." *Psychology and Aging* 20, no. 4 (2005): 542–53.

Carstensen, L. L., D. M. Isaacowitz, and S. T. Charles. "Taking Time Seriously: A Theory of Socioemotional Selectivity." *American Psychologist* 54 (1999): 165–81.

Charles, Susan Turk, Chandra A. Reynolds, and Margaret Gatz. "Age-Related Differences and Change in Positive and Negative Affect over 23 Years." *Journal of Personality and Social Psychology* 80, no. 1 (2001): 136–51.

4 Experience. Judgment. Wisdom.

In addition to interviews, I referred to or used as background the following studies, articles, and books:

Hess, Thomas M., Nicole L. Osowski, and Christine M. Leclerc. "Age and Experience Influences on the Complexity of Social Inferences." *Psychology and Aging* 20, no. 3 (2005): 447–59.

Hess, Thomas M. "Adaptive Aspects of Social Cognitive Functioning in Adulthood: Age-Related Goal and Knowledge Influences." *Social Cognition* 24, no. 3 (2006): 279–309.

Charles, Susan Turk. "Viewing Injustice: Greater Emotion Heterogeneity with Age." *Psychology and Aging* 20, no. 1 (2005): 159–64.

Goldberg, Elkhonon. *The Wisdom Paradox*. New York: Penguin Group (USA) Inc., 2005.

Bartzokis, George, Mace Beckson, Po H. Lu, Keith H. Nuechterlein, Nancy Edwards, and Jim Mintz. "Age-Related Changes in Frontal and Temporal Lobe Volumes in Men." *Archive of General Psychiatry* 58 (2001): 461–65.

Benes, F. M., M. Turtle, Y. Khan, and P. Farol. "Myelination of Key Relay Zone in the Hippocampal Formation Occurs in the Human Brain During Childhood, Adolescence, and Adulthood." *Archive of General Psychiatry* 51 (1994): 477–84.

Sowell, E., P. Thompson, C. J. Holmes, R. Batth, T. Jerrigan, and A. W. Toga. "Localizing Age-Related Changes in Brain Structure between Childhood and

Adolescence Using Statistical Parametric Mapping." *Neuroimage* 9 (1999): 587–97.

Giedd, J., J. Blumenthal, N. O. Jeffries, F. X. Castellanos, H. Liu, P. Zijdenbos, T. Paus, A. C. Evans, J. L. Rapoport. "Brain Development during Childhood and Adolescence in a Longitudinal MRI Study." *Nature Neuroscience* 2 (1999): 861–63.

Hall, Stephen S. "The Older-and-Wiser Hypothesis." *New York Times Magazine,* May 6, 2007, 58.

Ardelt, M., and G. E. Vaillant. "Wisdom as a Cognitive, Reflective and Affective Three-Dimensional Personality Characteristic." Paper presented at the Gerontological Society of America Annual Meeting, San Francisco, November 2007.

Raz, N., F. M. Gunning, D. Head, J. H. Dupuis, J. McQuain, S. D. Briggs, W. J. Loken, A. E. Thorton, and J. D. Acker. "Selective Aging of the Human Cerebral Cortex Observed In Vivo: Differential Vulnerability of the Prefrontal Gray Matter." *Cerebral Cortex* 7 (1997): 268–82.

Reistad-Long, Sara. "Older Brain Really May Be a Wiser Brain." *New York Times,* May 20, 2008, sec. F, p. 5.

Betts, Lisa R., Christopher P. Taylor, Allison B. Sekuler, and Patrick J. Bennett. "Aging Reduces Centre-surround Antagonism in Visual Motion Processing." *Neuron* 45 (2005): 361–66.

Reyna, V. F., and B. Kiernan. "The development of gist versus verbatim memory in sentence recognition." *Developmental Psychology* 30 (1994): 178–91.

Koutstaal, W., and D. L. Schacter. "Gist based false recognition of pictures in older and younger adults." *Journal of Memory and Language* 37 (1997): 555–83.

Reyna, V. F., and C. J. Brainerd. "What Theories of Memory Tell Us About the Brain." Plenary address at the 112th annual convention of the American Psychological Association, Honolulu, 2004.

Some material in this chapter also came from a three-day conference called the "Summit on Cognitive Aging" in Washington, D.C., October 10–12, 2007. Organized by the National Institute on Aging, it brought together for the first time psychologists, neuroscientists, nutritionists, geneticists, and animal researchers.

5 *The Middle in Motion*

In addition to interviews, I relied on the following:

Wahl, Hans-Werner, and Andreas Kruse. "Historical Perspectives of Middle Age Within the Life Span." Chap. 1 in *Middle Adulthood,* edited by Sherry Willis and Mike Martin. Thousand Oaks, CA: Sage Publications, Inc., 2005.

Moen, Phyllis, and Elaine Wethington. "Midlife Development in a Life Course Context." Chap. 1 in *Life in the Middle,* edited by Sherry Willis and James D. Reid. San Diego, CA: Academic Press, 1999.

"The MIDUS National Survey: An Overview." In *How Healthy Are We?: A National Study of Well-Being at Midlife,* O. G., Brim, C. D. Ryff, and R. C. Kessler., Chicago: University of Chicago Press, 2004.

Helson, Ravenna, and Christopher J. Soto. "Up and Down in Middle Age: Monotonic and Nonmonotonic Changes in Roles, Status, and Personality." *Journal of Personality and Social Psychology* 89, no. 2 (2005): 194–204.

Helson, R., C. Jones, and V.S.Y. Kwan. "Personality Change Over 40 Years of Adulthood." *Journal of Personality and Social Psychology* 83 (2002): 752–66.

Wink, Paul, and Ravenna Helson. "Practical and Transcendent Wisdom: Their Nature and Some Longitudinal Findings." *Journal of Adult Development* 4, no. 1 (1997).

Mroczek, Daniel K., and Avron Spiro III. "Change in Life Satisfaction during Adulthood: Findings from the Veterans Affairs Normative Aging Study." *Journal of Personality and Social Psychology* 88, no. 1 (2005): 189–202.

Levinson, Daniel J., Charlotte N. Darrow, Edward B. Klein, Maria H. Levinson, and Braxton McKee. *The Seasons of a Man's Life.* New York: Ballantine Books, 1978.

Fingerman, Karen L., Pei-Chun Chen, Elizabeth Hay, Kelly E. Cichy, and Eva S. Lefkowitz. "Ambivalent Relations in the Parent and Offspring Relationship." *Journal of Gerontology* 61B, no. 3 (2006): 152–60.

Fingerman, Karen L. "'We Had a Nice Little Chat': Age and Generational Differences in Mothers' and Daughters' Descriptions of Enjoyable Visits." *Journal of Gerontology: Psychological Sciences* 55B (2000): 5–106.

Mroczek, Daniel K., and Christian M. Kolarz. "The Effect of Age on Positive and Negative Affect: A Developmental Perspective on Happiness." *Journal of Personality and Social Psychology* 75, no. 5 (1998): 1333–49.

Deykin, Eva Y., Shirley Jacobson, Gerald Klerman, and Maida Solomon. "The Empty Nest: Psychological Aspects of Conflict between Depressed Women and Their Grown Children." *American Journal of Psychiatry* (1966): 1422–26.

"The 'New' Pat Nixon." *Ladies' Home Journal,* February 1962, 124–25.

Bedford, V. H. "Sibling Relationships in Middle Adulthood and Old Age." In *Handbook on Aging and the Family,* edited by R. Blieszner and V. H. Bedford, 201–22. Westport, CT: Greenwood, 1995.

Bedford, V. H. "Sibling Relationship Troubles and Well-Being in Middle and Old Age," *Family Relations* 47 (1998): 369–76.

Bedford, V. H. "Ambivalence in Adult Sibling Relationships," *Journal of Family Issues* 10, no. 2 (1989): 221–24.

Goode, Erica. "New Study Finds Middle Age Is Prime of Life." *New York Times,* February 16, 1999, sec. F, 6.

Clay, Rebecca A. "Researchers Replace Midlife Myths with Facts." APA 324, April 2003.

Gallagher, Winifred. "Midlife Myths." *Atlantic Monthly,* May 1993, 51–68.

Rasky, Susan F. "Corporate Psychologist: Elliott Jaques, His ideas on Work Take Hold." *New York Times,* February 17, 1985, sec. 3, p. 8.

Lavietes, Stuart. "Elliott Jaques, 86, Scientist Who Coined 'Midlife Crisis.'" *New York Times,* March 17, 2003, sec. B, obituary, 7.

Updike, John. *Rabbit Redux.* New York: Ballantine Books, 1971.

Sheehy, Gail. *Passages.* New York: Bantam Books, 1974.

Chew, Peter. *The Inner World of the Middle-Aged Man.* New York: Macmillan Publishing Co, Inc., 1976.

6 What Changes with Time

This chapter was based on extensive interviews with dozens of scientists investigating the aging brain, as well as findings from a number of ground-breaking studies. The main studies and articles I used:

Burke, Deborah M., Donald G. Mackay, Joanna S. Worthley, and Elizabeth

Wade. "On the Tip of the Tongue: What Causes Word Finding Failures in Young and Older Adults." *Journal of Memory and Language* 30 (1991): 542–79.

James, Lori E., and Deborah M. Burke. "Phonological Priming Effects on Word Retrieval and Tip-of-the-Tongue Experience in Young and Older Adults." *Journal of Experimental Psychology, Learning, Memory and Cognition* 26 (2001): 1378–91.

Burke, Deborah M., Jill Kester Locantore, Ayda A. Austin, and Bryan Chae. "Cherry Pit Primes Brad Pitt, Homophone Priming Effects on Young and Older Adults' Production of Proper names." *Psychological Science* 15, no. 3 (2004): 164–70.

Burke, Deborah M., and Meredith A. Shafto. "Language and Aging." Chapter to appear in *The Handbook of Aging and Cognition,* edited by F. I.M. Craik and T. Salthouse. London: Taylor and Francis, forthcoming.

Warner, Judith. "A Hole in the Head." *New York Times,* November 25, 2007. http://warner.blogs.nytimes.com/2007/11/22/a-hole-in-the-head.

Gazzaley, Adam, Jeffrey W. Cooney, Jesse Rissman, and Mark D. Esposito. "Top-Down Suppression Deficit Underlies Working Memory Impairment in Normal Aging." *Nature Neuroscience* 8, no. 10 (2005): 1298–1300.

Gazzaley, Adam, and Mark D'Esposito. "Top-Down Modulation and Normal Aging." *Annals of New York Academy of Sciences* 1097 (2007): 67–83.

Grady, Cheryl L., Mellanie V. Springer, Donaya Hongwanishkul, Anthony R. McIntosh, and Gordon Winocur. "Age-related Changes in Brain Activity Across the Adult Lifespan." *Journal of Cognitive Neuroscience* 18 (2006): 227–41.

Moore, Tara L., Ronald J. Killiany, James G. Herndon, Douglas L. Rosene, and Mark B. Moss. "Executive System Dysfunction Occurs as Early as Middle-Age in the Rhesus Monkey." *Neurobiology of Aging* 27 (2006): 1484–93.

Moore, Tara L., Ronald J. Killiany, James G. Herndon, Douglas L. Rosene, and Mark B. Moss. "Impairment in Abstraction and Set-Shifting in Aged Rhesus Monkey." *Neurobiology of Aging* 24, no. 1 (January-February 2003): 125–34.

Sowell, Elizabeth R., Bradley S. Peterson, Paul M. Thompson, Suzanne E. Welcome, Amy L. Henkenius, and Arthur W. Toga. "Mapping Cortical Change across the Human Life Span." *Nature Neuroscience* 6, no. 3 (March 2003): 309–15.

Begley, Sharon. "The Upside of Aging." *Wall Street Journal,* February 16, 2007.

Hedden, Trey, and John D. E. Gabrieli. "Insights into the Aging Mind: A View from Cognitive Neuroscience." *Nature Reviews Neuroscience* 5 (2004): 87–96.

Raz, Naftali, and Karen M. Rodrigue. "Differential Aging of the Brain: Patterns, Cognitive Correlates and Modifiers." *Neuroscience and Biobehavioral Reviews* 30 (2006): 730–48.

Raz, N., U. Lindenberger, K. M. Rodrigue, K. M. Kennedy, D. Head. A. Williamson, C. Dahle, D. Gerstorf , and J. D. Acker. "Regional Brain Changes in Aging, Healthy Adults: General Trends, Individual Differences and Modifiers." *Cerebral Cortex* 15 (2005): 1676–89.

Craik, F.I.M., and T. A. Salthouse, eds. "Aging of the Brain and Its Impact on Cognitive Performance: Integration of Structural and Functional Findings." In *Handbook of Aging and Cognition* II, 1–90. Mahwah, NJ: Erlbaum, 2000.

Kim, Sunghan, Lynn Hasher, and Rose T. Zacks. "Aging and a Benefit of Distractibility." *Psychological Bulletin Review* 2 (April 14, 2007): 301–305.

Healey, M. Karl, Karen L. Campbell, and Lynn Hasher. "Cognitive Aging and Increased Distractibility: Costs and Potential Benefits." *Progress in Brain Research* 169 (2008): 353–63.

Some material in this chapter also came from the "Summit on Cognitive Aging" in Washington, D.C., October 10–12, 2007.

7 Two Brains Are Better Than One

Cabeza, Roberto, Cheryl L. Grady, Lars Nyberg, Anthony R. McIntosh, Endel Tulving, Shitj Kapur, Janine M. Jennings, Sylvian Houle, and Fergus I. M. Craik. "Age-Related Differences in Neural Activity During Memory Encoding and Retrieval: A Positron Emission Tomography Study." *Journal of Neuroscience* 17, no. 10 (January 1, 1997): 391–400.

Cabeza, Roberto, Nicole D. Anderson, Jill K. Locantore, and Anthony R. McIntosh. "Aging Gracefully: Compensatory Brain Activity in High-Performing Older Adults." *NeuroImage* 17 (2002): 1394–1402.

Cabeza, Roberto. "Hemispheric Asymmetry Reduction in Older Adults: The HAROLD Model." *Psychology and Aging* 17, no. 1 (2002): 85–100.

Cabeza, Roberto, Sander M. Daselaar, Florin Dolcos, Steven E. Prince,

Matthew Budde, and Lars Nyberg. "Task-Independent and Task-Specific Age Effects on Brain Activity During Working Memory, Visual Attention and Episodic Retrieval." *Cerebral Cortex* 14, no. 4 (2004): 364–375.

Reuter-Lorenz, Patricia A., and Cindy Lustig. "Brain Aging: Reorganizing Discoveries About the Aging Mind." *Current Opinion in Neurobiology* 15 (2005): 245–51.

Cohen, Gene. "The Myth of the Midlife Crisis." *Newsweek*, January 16, 2006.

Park, D. C., T. A. Polk, R. Park, M. Menear, A. Savage, and M. R. Smith. "Aging Reduces Neural Specialization in Ventral Visual Cortex." *Proceedings of the National Academy of Sciences (PNAS)* 101(35) 13091–13095, 2004.

Park, Denise, and Patricia Reuter-Lorenz. "The Adaptive Brain: Aging and Neurocognitive Scaffolding." *Annual Reviews Psychology* 60, no. 21.1 (2009): 21–24.

Cohen, Gene D. *The Mature Mind*. New York: Basic Books, 2005.

Park, D. C., R. C. Welsh, C. Marshuetz, A. H. Gutchess, J. Mikels, and T. A. Polk. "Working Memory for Complex Scenes: Age Differences in Frontal and Hippocampal Activations." *Journal of Cognitive Neuroscience* 15, no. 8 (2003): 1122–34.

Reuter-Lorenz, P. A., and K. Cappell. "Neurocognitive Aging and the Compensation Hypothesis." *Current Directions in Psychological Science* 18, no. 3 (2008): 177–81.

Reuter-Lorenz, P. A., J. Jonides, E. E. Smith, A. Hartley, A. Miller, and C. Marshuetz. "Age Differences in the Frontal Lateralization of Verbal and Spatial Working Memory Revealed by PET." *Journal of Cognitive Neuroscience* 12, no. 1 (2000): 174–87.

Lu, Tao, Ying Pan, Shyan-Yuan Kao, Cheng Li, Isaac Kohane, Jennifer Chan, and Bruce A. Yankner. "Gene Regulation and DNA Damage in the Aging Human Brain." *Nature* 429 (June 24, 2000): 883–91.

Yankner, Bruce. "The Aging Brain: Gene Expression in Middle Age May Hold Clues to Cognitive Decline." *On the Brain: The Harvard Mahoney Neuroscience Institute Letter* 12, no. 2 (Spring 2006): 2–3.

Some material in this chapter also came from the "Summit on Cognitive Aging" in Washington, D.C., October 10–12, 2007.

8 Extra Brainpower

Snowdon, David. *Aging with Grace*. New York: Bantam, 2001.

Mortimer, J. A. "Brain Reserve and the Clinical Expression of Alzheimer's Disease." *Geriatrics* 52 (1993): 50–53.

Snowdon, David A. "Healthy Aging and Dementia: Findings from the Nun Study." *Annals of Internal Medicine* 139, no. 5 (December 2, 2003): 450–54.

Snowdon, D. A., L. H. Greiner, and W. R. Markesbery. "Linguistic Ability in Early Life and the Neuropathology of Alzheimer's Disease and Cerebrovascular Disease." Findings from the Nun Study, *Annals of the New York Academy of Sciences* 903 (2000): 34–38.

Melton, Lisa. "Use it, Don't Lose it." *New Scientist*, December 17, 2005, 32–35.

Katzman, Robert, Robert Terry, Richard DeTeresa, Theodore Brown, Peter Davies, Paula Fuld, Xiong Renbing, and Arthur Peck. "Clinical, Pathological and Neurochemical Changes in Dementia: A Subgroup with Preserved Mental Status and Numerous Neocortical Plaques." *Annals of Neurology* 23 (1988): 138–34.

Katzman, R., M. Aronson, P. Fuld, et al. "Development of Dementing Illnesses in an 80-year-old Volunteer Cohort." *Annals of Neurology* 25 (1989): 317–24.

Katzman, R. "Education and the Prevalence of Dementia and Alzheimer's Disease." *Neurology* 43 (1993): 13–20.

Hill, L. R., M. R. Klauber, D. P. Salmon, E.S.H. Yu, W. T. Liu, M. Zhang, and R. Katzman. "Functional Status, Education and the Diagnosis of Dementia in The Shanghai Survey." *Neurology* 43 (1993): 138–45.

Zhang, Mingyaun, Robert Katzman, David Salmon, Hua Jin, Guojun Cai, Zhengyu Wang, Guangya Qu. "The Prevalence of Dementia and Alzheimer's Disease in Shanghai, China: Impact of Age, Gender and Education." *Annals of Neurology* 27, no. 4 (1990): 428–37.

Kolata, Gina. "A Surprising Secret to a Long Life: Stay in School." *New York Times*, January 3, 2007.

Stern, Yaakov, Barry Gurland, Thomas Tatemichi, Ming Xin Tang, David Wilder, and Richard Mayeux. "Influence of Education and Occupation on the Incidence of Alzheimer's Disease." *Journal of the American Medical Association* 271, no. 13 (April 6, 1994): 1004, 1010.

Scarmeas, N., S. M. Albert, J. J. Manley, and Y. Stern. "Education and Rates of Cognitive Decline in Incident Alzheimer's Disease." *Journal of Neurology, Neurosurgery and Psychiatry* 77 (2005): 308–16.

Stern, Y., G. E. Alexander, I. Prohovnik, and R. Mayeux. "Inverse Relationship between Education and Parietotemporal Perfusion Deficit in Alzheimer's Disease. *Annals of Neurology* 32 (1992): 371–75.

Alexander, G. E., M. L. Furey, C. L. Grady, et al. "Association of Premorbid Function with Cerebral Metabolism in Alzheimer's Disease: Implications for the Reserve Hypotheses." *American Journal of Psychiatry* 154 (1997): 165–72.

Stern, Y., G. E. Alexander, I. Prohovnik, et al. "Relationship between Lifetime Occupation and Parietal Flow: Implications for a Reserve Against Alzheimer's Disease Pathology." *Neurology* 45 (1995): 55–60.

Stern, Yaakov, Nikolaos Scarmeas, and Chistian Habeck. "Imaging Cognitive Reserve." *International Journal of Psychology* 39, no. 1 (2004): 18–36.

Scarmeas, Nikolaos, Eric Zarahn, Karen Anderson, Lawrence S. Honig, Aileen Park, John Hilton, Joseph Flynn, Harold A. Sackeim, and Yaakov Stern. "Cognitive Reserve—Mediated Modulation of Positron Emission Tomographic Activations during Memory Tasks in Alzheimer Disease." *Archives of Neurology* 61 (January 2004): 73–78.

Scarmeas, Nikolaos, Eric Zarahn, Karen Anderson, Chistian G. Habeck, John Hilton, Joseph Flynn, Karen S. Marder, et al. "Association of Life Activities with Cerebral Blood Flow in Alzheimer's Disease." *Archives of Neurology* 60 (March 2003): 365–69.

Scarmeas, N., G. Levy, M. X. Tang, J. Manly, and Y. Stern. "Incidence of Leisure Activity on the Incidence of Alzheimer's Disease." *Neurology,* December 2001.

Scarmeas, Nikolaos, and Yaakov Stern. "Cognitive Reserve and Lifestyle." *Journal of Clinical and Experimental Neuropsychology* 25, no. 5 (2003): 625–33.

Stern, Yaakov, ed. *Cognitive Reserve.* New York: Taylor and Francis, 2007.

Wilson, Robert S., Carlos F. Mendes de Leon, Lisa L. Barnes, Julie A. Schneider, Julia L. Bienias, Denis A. Evans, and David A. Bennett. "Participation in Cognitively Stimulating Activities and the Risk of Incident Alzheimer's Disease." *Journal of the American Medical Association* 287, no. 6 (February 13, 2002): 742–48.

Wilson, R. S., Y. Li, N. T. Aggarwal, L. L. Barnes, J. J. McCann, D. W. Gilley, et al. "Education and the Course of Cognitive Decline in Alzheimer's Disease." *Neurology* 63 (2004): 1198–1202.

Bleeker, M. L., D. P. Ford, C. G. Vaughan, and K. N. Lindgren. "Impact of Cognitive Reserve on the Relationship of Lead Exposure and Neurobehavioral Performance." *Neurology* 69 (2007): 470–76.

Kesler, Shelli R., Heather F. Adams, Christine M. Blasey, and Erin D. Bigler. "Premorbid Intellection Functioning, Education and Brain Size in Traumatic Brain Injury: An Investigation of the Cognitive Reserve Hypothesis." *Applied Neuropsychology* 10, no. 3 (2003): 153–62.

9 *Keep Moving and Keep Your Wits*

Pereira, Ana C., Dan E. Huddleston, Adam M. Brickman, Alexander A. Sosunov, Rene Hen, Guy M. McKhann, Richard Sloan, Fred H. Gage, Truman R. Brown, and Scott A. Small. "An In Vivo Correlate of Exercise-Induced Neurogenesis in the Adult Dentate Gyrus." *PNAS* 104, no. 13 (March 27, 2007): 5638–43.

Small, Scott A., Monica K. Chawla, Michael Buonocore, Peter R. Rapp, and Carol A. Barnes. "Imaging Correlates of Brain Function in Monkeys and Rats Isolates a Hippocampal Subregion Differentially Vulnerable to Aging." *PNAS* 101, no. 18 (May 4, 2004): 7181–86.

Small, Scott A., Wei Yann Tsai, Robert DeLaPaz, Richard Mayeux, and Yaakov Stern. "Imaging Hippocampal Function Across the Human Life Span: Is Memory Decline Normal or Not?" *Annals of Neurology* 51 (2002): 290–95.

Sloan, Richard P., Peter A. Shapiro, Ronald E. DeMeersman, Paula S. McKinley, Keven J. Tracey, Iordan Slavov, Yixin Fank, and Pamela D. Flood. "Aerobic Exercise Attenuates Inducible TNF Production in Humans." *Journal of Applied Physiology* 103 (2007): 1007–11.

Gage, Fred H. "Brain, Repair Yourself." *Scientific American,* September 2003, 47–53.

Begley, Sharon. *Train Your Mind, Change Your Brain.* New York: Ballantine Books, 2007.

Reynolds, Gretchen. "Lobes of Steel." *New York Times, Play* magazine, August 19, 2007.

Vastag, Brian. "Brain Gain." *Science News* 171 (June 16, 2007): 376–80.

van Praag, Henriette, Brian R. Christie, Terrence J. Sejnowski, and Fred H..

Gage. "Running Enhances Neurogenesis, Learning and Long-Term Potentiation in Mice." *PNAS* 96, no. 23 (November 9, 1999): 13427–31.

Eriksson, Peter S., Ekaterina Perfilieva, Thomas Bjork-Eriksson, Ann-Marie Alborn, Claes Nordborg, Daniel A. Peterson, and Fred H. Gage. "Neurogenesis in the Adult Human Hippocampus." *Nature Medicine* 4 (1998): 1313–17.

van Praag, H., G. Kempermann, F. H. Gage. "Running Increases Cell Proliferation and Neurogenesis in the Adult Mouse Dentate Gyrus." *Nature Neuroscience* 2, no. 3 (March 1999): 266–70.

van Praag, Henriette, Tiffany Shubert, Chunmei Zhao, and Fred H. Gage. "Exercise Enhances Learning and Hippocampal Neurogenesis in Aged Mice." *Journal of Neuroscience* 25, no. 38 (September 21, 2005): 8680–85.

Gould, Elizabeth, Anna Beylin, Patima Tanapat, Alison Reeves, and Tracey J. Shors. "Learning Enhances Adult Neurogenesis in the Hippocampal Formation." *Nature Neuroscience* 2 (1999): 260–65.

Carmichael, Mary. "Stronger, Faster, Smarter." *Newsweek,* March 26, 2007.

Colcombe, Stanley J., Kirk I. Erickson, Paige E. Scalf, Jenny S. Kim, Ruchika Prakash, Edward McAuley, Steriani Elavsky, David X. Marquez, Liang Hu, and Arthur K. Kramer. "Aerobic Exercise Training Increases Brain Volume in Aging Humans." *Journals of Gerontology Series A: Biological Sciences and Medical Sciences* 61 (2006): 1166–70.

Colcombe, Stanley J., Kirk I. Erickson, Naftali Raz, Andrew G. Webb, Neal J. Cohen, Edward McAuley, and Arthur F. Kramer. "Aerobic Fitness Reduces Brain Tissue Loss in Aging Humans." *Journals of Gerontology* 58 (2003): M176–80.

Kramer, A. F., S. Hahn, N. J. Cohen, M. R. Banich, E. McAuley, C. R. Harrison, et al. "Aging, Fitness and Neurocognitive Function." *Nature* 400 (1999): 418–19.

Churchill, James D., Roberto Galvez, Stanley Colcombe, Rodney A. Swain, Arthur F. Kramer, and William T. Greenough. "Exercise, Experience and the Aging Brain." *Neurobiology of Aging* 23 (2002): 941-55.

"OHSU Researchers Study Physical and Mental Impacts of Exercise on the Brain." Findings presented at the November 6, 2003, annual meeting of the Society for Neuroscience, New Orleans.

Gage, Fred, and Janet Wiles. "Newborn Brain Cells 'Time-Stamp' Memories." *Neuron,* January 29, 2009, 187–202.

10 Food for Thought

Pollack, Andrew. "Glaxo Says Wine May Fight Aging." *New York Times,* April 23, 2008, sec. C, 11.

Wade, Nicholas. "New Hints Seen That Red Wine May Slow Aging." *New York Times,* June 4, 2008, sec. A, 1.

Bhagavad Gita. Translated by Stephen Mitchell. New York: Three Rivers Press, 2000.

Fontana, Luigi, and Samuel Klein Samuel. "Aging, Adiposity and Calorie Restriction." *Journal of the American Medical Association* 297 (2007): 986–94.

Kolata, Gina. "Low-Fat Diet Does Not Cut Health Risks." *New York Times,* February 8, 2006, sec. A, 1.

Kolata, Gina. "Maybe You're Not What You Eat." *New York Times,* February 14, 2006, sec. F, 1.

Taubes, Gary. "Do We Really Know What Makes Us Healthy?" *New York Times Magazine,* September 16, 2007, 52.

Pollan, Michael. *In Defense of Food.* New York: Penguin Press, 2008.

Pollan, Michael. "Unhappy Meals." *New York Times Magazine,* January 28, 2007, 38.

Milgram, N. W., E. Head, S. C. Zicker, C. J. Ikeda-Douglas, H. Murphey, B. Muggenburg, C. Siwak, D. Tapp, and C. W. Cotman. "Learning Ability in Aged Beagle Dogs is Preserved by Behavioral Enrichment and Dietary Fortification: A Two-Year Longitudinal Study." *Neurobiology of Aging* 26 (2005): 77–90.

Bakalar, Nicholas. "It Can Be Done: Scientists Teach Old Dogs New Tricks." *New York Times,* January 25, 2005, sec. F (Science), 1.

Morrison, John H., Roberta D. Brinton, Peter J. Schmidt, and Andrea C. Gore. "Estrogen, Menopause, and the Aging Brain: How Basic Neuroscience Can Inform Hormone Therapy in Women." *Journal of Neuroscience* 26, no. 41 (October 11, 200): 10332–48.

Porter, Peggy. "Westernizing Women's Risks? Breast Cancer in Lower-Income Countries." *New England Journal of Medicine,* Perspective 358 (January 17, 2008): 3.

Bakalar, Nicholas. "Study Critiques Antioxidant Supplements." *New York Times,* April 29, 2008, sec. F.

Wade, Nicholas. "Pill to Extend Life? Don't Dismiss the Notion Too Quickly." *New York Times,* September 22, 2000, sec. A, 20.

Cartford, M. Claire, Carmelina Gemma, and Paula C. Bickford. "Eighteen-Month-Old Fischer 344 Rats Fed a Spinach-Enriched Diet Show Improved Delay Classical Eyeblink Conditioning and Reduced Expression of Tumor Necrosis Factor in the Cerebellum." *Journal of Neuroscience* 14 (July 15, 2002): 5813–16.

Gemma, Carmelina, Michael H. Mesches, Boris Sepesi, Kevin Choo, Douglas B. Holmes, and Paula C. Bickford. "Diets Enriched in Food with High Antioxidant Activity Reverse Age-Induced Decreases in Cerebellar Adrenergic Function and Increases in Proinflammatory Cytokines." *Journal of Neuroscience* 14 (July 15, 2002): 6114–20.

Stromberg, I., C. Gemma, J. Vila, and P. C. Bickford. "Blueberry and Spirulina-Enriched Diets Enhance Striatal Dopamine Recovery and Induce a Rapid, Transient Microglia Activation after Injury of the Rat Nigrostriatal Dopamine System." *Experimental Neurology* 196 (2005): 298–307.

Joseph, J. A., B. Hale-Shukitt, N. A. Denisova, D. Bielinski, A. Martin, J. J. McEwen, and P. C. Bickford. "Reversals of Age-Related Declines in Neuronal Signal Transduction Cognitive and Motor Behavioral Deficits with Blueberry, Spinach or Strawberry Dietary Supplements." *Journal of Neuroscience* 19 (1999): 8144–21.

Joseph, J. A., B. Hale-Shukitt, N. A. Denisova, R. L. Prior, G. Cao, A. Martin, G. Taglialatela, and P. C. Bickford. "Long-Term Dietary Strawberry, Spinach or Vitamin E Supplementation Retards the Onset of Age-Related Neuronal Signal-Transduction and Cognitive Behavioral Deficits." *Journal of Neuroscience* 18 (1998): 8047–55.

Joseph, J. A., G. Arendash, M. Gordon, D. Diamond, B. Hale Shukitt, and D. Morgan. "Blueberry Supplementation Enhances Signaling and Prevents Behavioral Deficits in an Alzheimer Disease Model." *Nutrition Neuroscience* 6 (2003): 153–62.

"Smart Drugs." Medicine. *The Economist,* May 24, 2008.

"All On the Mind." Science and Technology. *The Economist,* May 24, 2008.

Sahakian, Barbara, and Sharon Morein-Zamir. "Professor's Little Helper." *Nature* 450 (December 20, 2007): 1157–59.

Maher, Brendan. "Poll Results: Look Who's Doping." *Nature* 452, no. 10 (April 2008).

"Brain Boosting Drugs Hit the Faculty Lounge." *The Chronicle of Higher Education*, December 20, 2007. http://chronicle.com/news/article/3673/brain-boosting-drugs-hit-the-faculty-lounge.

Carey, Benedict. "Smartening Up: Brain Enhancement is Wrong, Right?" *New York Times*, March 9, 2008, Week in Review.

Maswood, Navin, Jennifer Young, Edward Tilmont, Zhiming Zhang, Don M. Gash, Greg. A. Gerhardt, Richard Grondin, et al. "Caloric Restriction Increases Neurotrophic Factor Levels and Attenuates Neurochemical and Behavioral Deficits in a Primate Model of Parkinson's Disease." *PNAS* 101, no. 42 (2004): 18171–76.

McGlothin, Paul, and Meredith Averill. *The CR Way*. New York: Harper Collins Publishers, 2007.

Fontana, Luigi, and Samuel Klein. "Aging, Adiposity, and Calorie Restriction." *Journal of the American Medical Association* 297, no. 9 (March 7, 2007): 986–94.

Scarmeas, Nikolaos, Yaakov Stern, Richard Mayeux, Jennifer J. Manly, Nicole Schupf, and Jose A. Luchsinger. "Mediterranean Diet and Mild Cognitive Impairment." *Archives of Neurology* 66, no. 2 (February 2009): 216–25.

Nagourney, Eric. "Tie between Obesity and Dementia Is Discovered." *New York Times*, November 1, 2005, sec. F, 6.

Grady, Denise. "Link Between Diabetes and Alzheimer's Deepens." *New York Times*, July 17, 2006, sec. A, 15.

Caryn Rabin, Roni. "Blood Sugar Control Linked to Memory Decline, Study Says." *New York Times*, January 1, 2009.

Pearson, Kevin J., Joseph A. Baur, Kaitlyn N. Lewis, Leonid Peshkin, et al. "Resveratrol Found to Improve Health, but Not Longevity, in Aging Mice on Standard Diet." *Cell Metabolism*, July 3, 2008.

11 The Brain Gym

Liu, Linda, and Denise Park. "Imagination Helps Older People Remember to Comply with Medical Advice." *National Institutes of Health News*, June 4, 2004.

Park, Denise C., Angela H. Gutchess, Michelle L. Meade, and Elizabeth A. L. Stine-Morrow. "Improving Cognitive Function in Older Adults: Nontraditional Approaches." *Journal of Gerontology*, Series B, 62B (special issue) (2007): 45–52.

Liu, L. L., and D. C. Park. "Aging and Medical Adherence." *Psychology and Aging* 19 (2004): 318–25.

"New Research on Aging and Cognitive Training Presented at The Gerontological Society of America's Annual Meeting." The Gerontological Society of America, November 19, 2007. IMPACT study by Elizabeth Zelinski.

Mahncke, Henry W., Bonnie B. Connor, Jed Appelman, Omar N. Ahsannuddin, Joseph L. Hardy, Richard A. Wood, Nicholas M. Joyce, Tania Boniske, Sharona M. Atkins, and Michael M. Merzenich. "Memory Enhancement in Healthy Older Adults Using a Brain Plasticity-Based Training Program: A Randomized, Controlled Study." *PNAS*, August 3, 2006, 12523–28.

Merzenich, Michael M. "Building Better Brains from Lab to Laptop." *Aging Today* XXVII, no. 1 (January-February 2006): 1043–1284.

George, Lianne. "The Secret to Not Losing Your Marbles." *MacLeans*, April 9, 2007.

Trafford, Abigail. "In the Key of What-Next." My Time, *Washington Post*, May 11, 2004, Health.

Fried, Linda P., Michelle C. Carlson, Marc Freedman, Kevin D. Frick, Thomas A. Glass, Joel Hill, Sylvia McGill, et al. "A Social Model for Health Promotion for an Aging Population: Initial Evidence on the Experience Corps Model." *Journal of Urban Health: Bulletin of the New York Academy of Medicine*, 81, no. 1, March 2004, 64–68.

Kolata, Gina. "Old but Not Frail: A Matter of Heart and Head." *New York Times*, October 5, 2006, sec. A, 1.

Levy, Becca. "Negative Stereotypes About Aging May Shorten Life." *Personality and Social Psychology*, August 2002. http://www/apa.org/releases/longevity.html.

Hess, Thomas M., Corinne Auman, Stanley J. Colcombe, and Tamara A. Rahhal. "The Impact of Stereotype Threat on Age Differences in Memory Performance." *Journal of Gerontology, Psychological Sciences* 58B, no. 1 (2003): 3–11.

"The Joy of Giving." *The Economist*, October 14, 2006, Science and Technology, 86. *PNAS*, October 12, 2006.

Schnall, Simone, et al. "Making It Less of an Up Hill Struggle." *Journal of Experimental Social Psychology*, June 14, 2008.

Reuter-Lorenz, P. A. "New Visions of the Aging Mind and Brain." *Trends in Cogntive Sciences* 6 (2002): 394–400.

Reuter-Lorenz, P. A., and C. Lustig. "Brain Aging: Reorganizing Discoveries About the Aging Mind." *Current Opinion in Neurobiology* 15 (2005): 245–51.

Salthouse, T. A. "The Processing-Speed Theory of Adult Age Differences in Cognition." *Psychological Review* 103 (1996): 403–28.

Salthouse, T. A. "Mental Exercise and Mental Aging: Evaluating the Validity of the Use It or Lose It Hypothesis." *Perspectives in Psychological Science* 1 (2006): 68–87.

Anderson, Nichole D., Patricia L. Ebert, Janine M. Jennings, Cheryl L. Grady, Roberto Cabeza, and Simon J. Graham. "Recollection and Familiarity-Based Memory in Healthy Aging and Amnestic Mild Cognitive Impairment." *Neuropsychology* (accepted for publication), July 30, 2007.

Ramachandran, Vilayanur S., and Lindsay M. Oberman. "Broken Mirrors." *Scientific American,* November 2006, 63–70.

Blakeslee, Sandra. "The Cells That Read Minds." *New York Times,* January 10, 2006, sec. F, 1.

Iacoboni, Marco. "Mirroring People." New York: Farrar, Straus and Giroux, 2008.

Some material in this chapter also came from the "Summit on Cognitive Aging" in Washington, D.C., October 10–12, 2007.

Epilogue

Block, Sandra, and Stephanie Armour. "Many Americans Retire Years Before They Want To." *USA Today,* July 10, 2006, 1.

Saletan, William. "Nerdo-Enhancement." *Slate,* April 10, 2008. http://www.slate.com/id/2188747/.

Cascio, Jamais. "Getting Smarter About Getting Older." Institute for Ethics and Emerging Technologies, July 7, 2005. http://ieet.org/index.php/IEET/more/cascio20050706.

"Global Aging." *Business Week,* January 31, 2005.

Marquez, Jessica. "Novel Ideas at Borders Lure Older Workers." *Workforce Management,* May 2005.

Deutsch, Claudia H. "A Longer Goodbye." *New York Times,* April 21, 2008.

"Why Population Aging Matters: A Global Perspective." National Institute on Aging, National Institutes of Health, U.S. Department of Health and Human Services, U.S. Department of State (2007): Special Section, 1.

Gawande, Atul. "The Way We Age Now." Annals of Medicine. *The New Yorker,* April 30, 2007.

Zelinski, Elizabeth, and Kerry P. Burnight. "Sixteen-Year Longitudinal and Time Lag Changes in Memory and Cognition in Older Adults." *Psychology and Aging* 12, no. 3 (1997): 503–13.

Brooks, David. "The Great Forgetting." *New York Times,* April 11, 2008, sec. A, 23.

Langa, Kenneth, et al. "Memory Loss and Other Cognitive Impairment Becoming Less Common in Older Americans." *Alzheimer's & Dementia* and *University of Michigan Health System,* February 18, 2008. http://www2.med.umich.edu/prmc/media/newsroom/details.cfm?ID=38.

Albert, Marilyn, and Guy McKhann. "The Aging Brain." Dana Alliance for Brain Initiatives, 2006. Pamphlet published by the Dana Alliance, Washington, D.C.

Lohr, Steve. "For a Good Retirement, Find Work, Good Luck." *New York Times,* June 22, 2008, Week in Review, 3.

Trafford, Abigail. "An Extra Ten and Young Again." My Time. *Washington Post,* September 18, 2007.

Shoven, John B. "New Age Thinking: Alternative Ways of Measuring Age, Their Relationship to Labor Force Participation, Government Politics and GDP." Working paper 13476, National Bureau of Economic Research, Cambridge, MA, October 2007, 1–19.

McFadden, Robert. "All 155 Aboard Safe as Crippled Jet Crash-Lands in Hudson." *New York Times,* January 16, 2009, sec. A, 1.

Wald, Matthew, and Al Baker. "Cockpit Tape Reveals Thumps, Engine Loss and a 'Mayday,'" *New York Times,* January 19, 2009, sec. A, 21.

Collins, Gail. "The Stump Theory." *New York Times,* February 12, 2009, sec. A, 35.

Index

Aberdeen, University of, 120
acetylcholine, 181
age discrimination, 192–94
aging:
    "brain-damage" model of, 92–93
    caloric restriction and, 160–67
    creativity and, 49–50, 89–90,
      98–99
    dementia and, *see* dementia
    education and, xxii, 111–17, 121,
      126, 145
    emotions and, 28–40, 41–42, 43,
      48, 56–58, 61, 64, 69–80
    genes and, xiv, 99–103, 121, 142,
      163
    inflammation and, 132, 142,
      149–53, 156, 167, 168
    loss equated with, 16–17
    memory and, *see* memory
    nutrition and, xxii–xxiii, 144–69,
      171
    pattern recognition and, 46–48
    self-image and, 186
    social interaction and, 184–85,
      186–87, 188–90
    vocabulary growth and, 14, 15,
      16, 71, 171
    worries about brain function and,
      3–11
    see also brain; middle age
"Aging and Emotional Memory:
    The Forgettable Nature of
    Negative Images for Older
    Adults" (Carstensen), 35
"Aging Gracefully: Compensatory
    Brain Activity in High-
    Performing Older Adults"
    (Cabeza), 96

Aging with Grace (Snowdon), 105
Agriculture Department, U.S.,
    151
air-traffic controllers, 20–21,
    26–27
Albert, Marilyn, 77, 81
Albert Einstein College of
    Medicine, 108
Alchemist, The (Coelho), 4
alfalfa sprouts, 151
Almeida, David M., 57
Altman, Joseph, 130
Alzheimer's disease, xxii, 10, 18,
    73, 81, 83–84
    cognitive reserve and, 105–7,
      108–13, 115–16, 117, 119, 157,
      176–77
    nutrition and, 142, 150, 152, 156,
      162, 168, 169
    social interaction and, 184
    see also dementia
American Journal of Psychiatry, 63
amygdala, 31–32, 37, 39, 198
Anderson, Nicole, 173–74
antidepressants, 154–55
anti-inflammatories, 148, 152–53,
    158
antioxidants, xxiii, 145, 146, 148,
    149–51, 152–53, 157, 158, 165,
    166, 167
    top foods for, 151
architecture, brain health and,
    185–86
Archives of Neurology, 168
Ardelt, Monika, 45–46
aspirin, 148, 158
atherosclerosis, 162
avocados, 151

221